THE ENVIRONMENT AS HAZARD

The Environment as Hazard

SECOND EDITION

Ian Burton
Robert W. Kates
Gilbert F. White

THE GUILFORD PRESS

New York / London

© 1993 The Guilford Press
A Division of Guilford Publications, Inc.
72 Spring Street, New York, NY 10012

Marketed and distributed outside North America by Longman
Group UK Limited

Printed in the United States of America

This book is printed on acid-free paper.

Last digit is print number: 9 8 7 6 5 4 3 2

Library of Congress Cataloging-in-Publication Data

Burton, Ian.
    The environment as hazard / Ian Burton, Robert W. Kates, and
Gilbert F. White. — 2nd ed.
        p.   cm.
    Includes bibliographical references and index.
    ISBN 0-89862-159-3 (pbk.)
    1. Natural disasters.   2. Air—Pollution.   3. Land use—
Environmental aspects.   4. Disaster relief planning.   I. Kates,
Robert William.   II. White, Gilbert F.   III. Title.
GB5014.B87 1993   363.3′4—dc20   92-33125
                                                          CIP

Permission to reprint an excerpt from "The Concept of Environmental Potential
as Exemplified by Tropical African Research" by P. Porter, in *Geography and
a Crowding World* (pp. 187–217), edited by W. Zelinsky, L. A. Kosinski, and M.
Prothero is gratefully acknowledged. Copyright 1970 by Oxford University Press.

Contents

Illustrations

Tables

Figures

Preface to the Second Edition

As we write this preface, recent drought in Africa, earthquake and brushfire in California, floods in China, tropical cyclones in Bangladesh and Florida, and volcanic eruptions in the Philippines have claimed large losses of property and lives and remind us again of the ongoing toll of natural hazards in a world preoccupied with major, often drastic, changes in society and economy. Continuing concerns about the consequences of global warming — the potential for new extremes in drought and flood — reinforce the continuing salience of this study of human adjustment to natural hazards. But requests for the reissue of the 1978 edition of this book have raised the question of whether knowledge of the situations and ideas it describes have changed to such an extent over the past 15 years to make it out of date. Our judgment is that the greater part of the text is still valid in 1993, but that any reprinting should incorporate a new introductory chapter and concluding statements of the major changes and challenges of the intervening years. A number of significant changes occurred in policy and program as well as theory, but they do not invalidate the basic thrust of the original volume.

We considered producing a more thorough revision of the earlier text but decided that would be difficult, and perhaps impossible, without the benefit of new assessments of the kind that formed the basis for *The Environment as Hazard*: the comparative study, under the auspices of the International Geographical Union, of diverse physical and social situations in a variety of countries — developing as well as developed (White, 1974), and the completion, with the support of the National Science Foundation, of a comprehensive *Assessment of Research on Natural Hazards* (White & Haas, 1975). In the interest of acquiring improved statistics, more incisive analysis, and more critical theory, new efforts should be undertaken without delay — a task clearly beyond the capacity of any individual author. But until that has been accomplished, there would seem to be justification for reissuing the earlier volume.

Within the new introductory and concluding chapters, we have sought to refer to major advances in the field since 1978, some of which involve new theory or criticism of our theory. Other advances involve new substantive studies, while others represent the formation of new institutions and public programs. In combination, these present a brief review of how research in an interdisciplinary field has changed in 15 years and of how research findings have been incorporated into action programs and public policy. This new text, is thus, the record of a field in rapid evolution. It is, further, an outline of urgent research needs, including a justification for updating in improved form the assessment effort of the early 1970s. As we said in our earlier preface, we hoped that the volume would foster a more searching examination of basic questions of methods and generalizations. That has happened as academic, government, and private efforts have proliferated. The time is ripe for a comprehensive assessment of where scholars and administrators stand today.

With the exception of the new introductory chapter that deals with current situations and retrospect, and the new concluding chapter on emerging synthesis, this volume varies with only minor changes from the 1978 edition.

We are grateful to the following people who during recent months have provided new information or helpful criticism of the revised text: Jeanne X. Kasperson, Roger Kasperson, Jolana Machalek, Mary Fran Myers, Pamela Showalter, and Billie L. Turner II.

IAN BURTON
ROBERT W. KATES
GILBERT F. WHITE

Preface

In one sense this volume is a summing up of what we think has been learned so far as to how individuals and social groups respond to extreme events in nature. In another sense it is a point of departure, as it contains conclusions that many of our colleagues believe should be re-examined, and uses methods that cry out for refinement. We are sensitive to those deficiencies, but have concluded that a rough syntheses at this time will foster rather than hinder more searching examination of the basic questions.

Although researches leading to this volume had their origin in earlier studies of floodplain adjustments and coastal flooding, they could not have expanded as they did after 1967 without the understanding support of two organizations. The National Science Foundation provided financial aid to Clark University and the University of Colorado (initially the University of Chicago) to carry out studies based at Worcester, Massachusetts, Boulder, Colorado, and at the University of Toronto in Canada. This made it possible to engage in cooperative studies with workers in other institutions, and Howard Hines of the Foundation gave discerning and laconic advice. The Commission on Man and Environment of the International Geographical Union (IGU) designated natural hazard investigation as one of its principal concerns, sponsored two meetings, and conducted a great deal of correspondence. The Commission has included, in addition to two of the authors, Professor David H. K. Amiran of the Hebrew University in Jerusalem, Academician I. P. Gerasimov of the Academy of Sciences of the Soviet Union, Professor Stanislaw Leszczycki of the Polish Academy of Sciences, Professor Takamasa Nakano of Tokyo Metropolitan University, and Professor R. Ramachandran of the University of Delhi. As Secretary-General of the IGU, Professor Chauncy D. Harris facilitated the Commission's work.

Sectors of related activity were supported in a variety of ways. In addition to financial aid from their home institutions, investigators received grants from several national foundations. UNESCO, in cooperation with the Hungarian Academy of Sciences, sponsored a seminar in Gödöllö in August 1970. Canadian activities were supported in part by the Canada Emergency Measures Organization and the National Advisory Committee for Water Resources Research. A conference in Calgary in July 1972 was sponsored by the Canadian National Advisory Committee for Geographical Research and the Government of Alberta. The Rockefeller Foundation, through its support of the Bureau of Resource Assessment and Land Use Planning at the University of Dar es Salaam, made possible much of the work in Tanzania. Resources for the Future, Inc., provided funds for a comparative investigation of air pollution in the United Kingdom.

Some of the papers comprising the program at Gödöllö have been printed in *Man and Environment,* edited by Marton Pecsi and Ferenc Probald (Budapest: Research Institute of Geography, Hungarian Academy of Sciences, 1974). The Calgary papers form a substantial part of the volume entitled *Natural Hazards: Local, National, Global,* edited by Gilbert F. White (New York: Oxford University Press, 1974).

Recent geographic study of natural hazards has engaged a broad array of people, and in drawing upon their findings we are keenly aware of our debt to them. To a remarkable degree their work has been international and collaborative. Through conferences and correspondence they have exchanged data as well as ideas involving both methods and conclusions. While it is difficult to name all those participating either in field investigations or in seminars and conferences in which research methods and findings have been reviewed, we can at least make mention of Robert L. A. Adams, Carlos Penaherrera del Aguila, E. M. Fournier d'Albe, Paula Archer, David G. Arey, Oliver M. Ashford, Andris Auliciems, Wilfrid Bach, Mary Barker, Duane D. Baumann, Leonard Berry, Jacquelyn Beyer, J. D. Billingsley, Mark Blacksell, Reuben H. Brooks, Vida Chapman, Rodney J. Cheatle, Earl Cook, Alexander J. Crosbie, Herbert Dupree, Fillmore C. F. Earney, Neil Ericksen, Leslie T. Foster, Phillip Frankland, James Gardner, James E. Goddard, Stephen M. Golant, A. P. Lino Grima, Andrew Guymer, J. Eugene Haas, Louis Hamill, Thomas D. Hankins, F. Kenneth Hare, Leslie R. Heathcote, Daya Hewapathirane, John Hewings, Kenneth Hewitt, Arild Holt-Jensen, James M. Houston, Nancy Hultquist, M. Aminul Islam, Edgar L. Jackson, Richard H. Jackson, A. Hersch Jacobs, David E. Kromm, Howard Kunreuther, Michel F. Lechat, Stanislaw Leszczycki, Lynne May,

Geoffrey R. McBoyle, Ross Mackay, Roger F. McLean, Bruce Mitchell, J. K. Mitchell, Norman Moline, Karen Moon, Tapan Mukerjee, Allan H. Murphy, Brian J. Murton, Ernst Neef, J. Gordon Nelson, Timothy O'Riordan, Masahiko Oya, Carlos Parra, Maurice Perry, Trevor W. Plumb, Philip Porter, Ferenc Probald, E. C. Relph, John Rooney, Clifford S. Russell, Rowan Rowntree, Justinian Rweymamu, Thomas F. Saarinen, Myra R. Schiff, T. Sekiguti, Christopher Saviour, W. R. Derrick Sewell, Lesley Sheehan, Shinzo Shimabukuro, Paul Simeon, Wendy Simpson, John H. Sims, Paul Slovic, Andras Szesztay, Chris Taylor, Keith Taylor, Stanko Vizjak, Geoffrey Wall, Robert M. Ward, Elizabeth Whitcombe, Anne U. White, Anne V. Whyte, Joe Whitney, Paul Wilkinson, Andrew W. Wilson, Ben Wisner, Bruce Wood, and T. Zvonkova. Their published contributions are indicated in appropriate notes below.

The location of the national studies generated by the IGU collaborative program is shown in the map appearing on pages 264 and 265. The investigations in those widely scattered places had a few methods in common but differed in the ways in which they reflected the cultural characteristics of place and investigator.

In the processing and analysis of research findings at Boulder and Toronto we were helped in important ways by Mary Barker, Nancy Simkowski Baumann, Brian A. Knowles, Neil Ericksen, Edgar Jackson, Donald L. Miller, Carlos Parra, Hazel Visvader, Anne U. White, and Paul Wilkinson. The graphics for this volume were the work of Nancy Fishman.

Secretarial and administrative assistance came from Paula Archer, Mary Bird, Judy Fukuhara, Sally Goodwillie, Jenny Haschak, Ronnie Mason, Jennie Racki, and Ingrid Shouldice. The burden of typing the manuscript in several versions fell on Jacque Myers. An earlier draft of the manuscript benefited from the careful editorial work of Lydia Burton.

The final version also incorporates many comments and suggestions of the following persons who kindly reviewed an earlier draft: Humberto Bravo A., Robert McC. Adams, James A. Anderson, David G. Arey, Oliver M. Ashford, Andris Auliciems, Duane D. Baumann, J. D. Billingsley, Reuben H. Brooks, Fillmore C. F. Earney, Neil Ericksen, Leslie T. Foster, Phillip Frankland, James Gardner, James E. Goddard, Stephen M. Golant, A. P. Lino Grima, Louis Hamill, R. Leslie Heathcote, Kenneth Hewitt, Janice R. Hutton, M. Aminul Islam, Edgar L. Jackson, David E. Kromm, Howard Kunreuther, Michel F. Lechat, Wendy Simpson Lewis, Roger F. McLean, J. K. Mitchell, Norman Moline, Ernst Neef,

Phil O'Keefe, Timothy O'Riordan, Ferenc Probald, Clifford S. Russell, Paul Slovic, Keith Taylor, Robert M. Ward, Anne U. White, Anne V. Whyte, Paul Wilkinson, Andrew W. Wilson, and Ben Wisner.

Our findings represent a combination and sieving of contributions from all these people. The task of advancing the analysis will fall on some of them and on others who we hope will turn their energies to a more penetrating examination of the environment as hazard.

IAN BURTON
ROBERT W. KATES
GILBERT F. WHITE

Is the Environment Becoming More Hazardous?

In a time of extraordinary human effort to live harmoniously in the natural world, the global toll from extreme events of nature is increasing. Loss in property from natural hazards is rising in most regions of the earth, and loss of life is continuing or increasing among many of the poor nations of the world.

In itself this paradox is not cause for exceptional concern; greater loss may denote larger populations, greater wealth, or social gain. With the population swelling at a rapid rate in much of the world, per capita loss could decrease or remain constant while the total increases. As floodplains are more intensively cropped, and new industry is attracted to the ocean shore, and more people find it possible to engage in winter sport, losses from flood, storm, and avalanche increase. These may, however, be overshadowed by the social benefits accruing.

It may well be that the ways in which humankind deploys its resources and technology in attempts to cope with extreme events of nature are inducing more rather than less damage and that the processes of rapid social change work in their own way to place more people at risk and make them more vulnerable.

To arrive at a truly accurate estimate of the degree to which the earth as the only home of the human race is becoming more hazardous would require a careful global count of loss in lives, health, property, and social functioning. It would also involve assessment of other costs that communities suffer in dealing with hazards: expenditure for flood-control works or hurricane warning services, or the extra expense of constructing office buildings that withstand earthquakes. It would take account of the benefits from these activities, and the opportunities—such as they are—to enrich life by means that would have been unavailable if people had not ventured onto hazard areas. Reaching a balanced estimate also requires a recognition that coping with one risk may open the door to another. Thus, the massive drought disaster in Sahelian Africa came upon the heels of apparent success in creating new sources of water supply which had been heralded as preventive measures but which had the opposite effect, as will be shown in the case of Tanzania.

In contrast to the global toll of hazard and the effort expended to prevent or reduce it, the social gain from a hazardous location has been assumed or described but never calculated. The benefit is substantial. People not only locate in areas of high, recurrent natural hazard; they survive and prosper there. Two catastrophes of the early 1970s illustrate this paradox. In magnitude of occurrence they were exceptional; in their social features they were commonplace.

BANGLADESH, 1970

The greatest natural disaster following World War II and prior to the Chinese earthquake of 1976 — the Bangladesh cyclone of November 1970 — was created by a severe tropical cyclone in the Bay of Bengal. On an average of 16 times over a 10-year period, winds exceeding 85 kilometers per hour lashed the low-lying coastal areas of the bay and were accompanied by wind-driven waves and storm surges.

The cyclone of 1970, which traversed the Indian Ocean, did not escape notice by the meteorological services of neighboring countries. Although it was identified on 9 November and tracked by satellite as it swept up the Bay of Bengal, no aircraft were available to determine its precise intensity; it was subsequently tracked by radar at Cox's Bazar (see Figure 1.1). At about 11 P.M. on 12 November it struck the coast — the peak of the storm surge nearly coinciding with high tide — and reached a maximum of almost 7 meters above normal high tide as it swept over the eastern coast of Bhola Island. The severe winds continued inland. By morning at least 225,000 people were dead, ripening crops worth $63 million were destroyed, and 280,000 head of cattle were swept away (Frank and Hussain, 1971, p. 439). No one knows the death toll for certain: in the emergency it was impossible to do more than bury the dead. As the central pressure of 950–960 millibars estimated for this storm indicates, it was not the most severe storm event ever to afflict this coast; a more severe storm occurred in 1876, and such a one might occur again.

BANGLADESH 1970

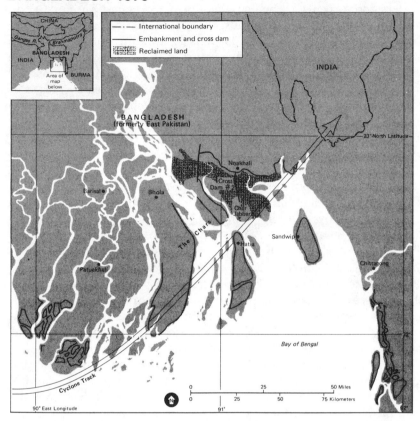

FIGURE 1.1. Reclaimed lands of Char Jabbar. The village of Char Jabbar was linked to the mainland of Noakhali District in 1964 by an 11-mile earthen embankment constructed to reclaim the river channel by deflecting flood and tidal flows to new channels. The embankment, Cross Dam No. 2, made possible the new settlements and intensified use of Char Jabbar (Islam, 1971). It was overtopped by the cyclone of November 12–13 but was not washed away. If adequate warning had been given, transportation prepared, and evacuation plans made in advance, the embankment could have served as an emergency evacuation route. Villagers reported that the warning of the cyclone, adequate in Dacca, never reached Char Jabbar. The telegraph office is 16 miles away, the radio day ended at 11:00 P.M., and there was much confusion over the type of radio warning issued. Motorized transportation was simply not available, families were reluctant to leave the household site or to subject women members to the public exposure of evacuation. Of an estimated 18,000 regular residents in the area, perhaps 6,000 died, mostly women and children, together with thousands of unknown migrant rice harvesters.

The causes of this disaster do not lie simply in the still poorly understood process by which the normally placid Indian Ocean transfers solar energy into atmospheric movement of a massive sort; land use can also be implicated. Densely inhabited offshore islands and low-lying coastal fringes form part of the deltas of many of the world's rivers, but the Ganges–Brahmaputra is the most intensively utilized. The government reclaimed land from the sea by means of sophisticated engineering and agricultural technology. New populations in large numbers were encouraged to occupy the resulting areas of great risk. Who was responsible for the ensuing disaster? Was the heightened risk justified by the benefits?

Some critics have argued that responsibility rests with those contributing to population pressure (Hardin, 1971); yet if a people is to be indicted for having too many children, other indictments necessarily follow. Of the farm land in Bangladesh, 20 percent is operated by tenants alone, and 40 percent by owners with tenants. The landlord's share of the harvest is normally one-half. While data are lacking, the landless laborers are estimated to comprise some 20–25 percent of the work force, and many of them flock to the shore in search of seasonal work or the opportunity to acquire land. What is responsible for such uneven land relations—a stagnant economy, a colonial inheritance, a foreign leadership or one preoccupied with politics of a kind that could erupt in civil strife?

It is facile to shift the blame onto the victims for having too many children or onto the social system for its great inequities. Engineering projects and warning channels are at least equally responsible because they are easier to manipulate than social systems. In 1970 many died who need not have, because of failure at governmental levels to weigh the full consequences of technological change on the indigenous system of livelihood. Public works requested in Karachi, designed in Holland, and financed in Washington provided means to stabilize the shifting islands (*chars*) of the delta and to keep out high tides and destructive salt. In so doing, they provided encouragement for settlement. What they did not provide was protection from cyclones or plans for evacuation.

Again, the triumph of satellite and radar technology enabled the meteorologists to give early warning, but this warning was not relayed by the radio station, which closed down at 11 P.M. Moreover, the message was confused by a newly adopted streamlined system of warning and blocked by officials high and low who failed to pass it on. The warning had little effect on farmers not organized to use it and fearful of forsaking home for unknown places of dubious safety.

The Dutch engineer, the international banker, the world meteorological expert, the Bengali radio station manager — all were, it may be assumed, conscientiously pursuing their tasks; yet ironically, disaster overcame their well-motivated desire to assist in the development process for a better life. They then joined in emergency relief and in reconstruction activities that were bound to persist for a long time before productivity could be fully regained.

Since all this occurred in one of the very poor nations of the world, some observers might explain it by the prevailing levels of social organization and technology. A somewhat comparable episode in the life of a rich and technically advanced nation can be taken from the same period of disaster history in the United States.

TROPICAL STORM AGNES, 1972

Another haunt of tropical cyclones — the Caribbean region — generated a huge disturbance in mid-June 1972. Agnes, as it was named by the international observation network, had taken clear shape off the coast of Yucatan by 15 June and was moving northward toward the western Florida shore of the United States. Land and ship reports and special observation planes formed an intricate system of tracking and prediction. The place and time of the storm's landfall was therefore forecast with reasonable accuracy. Although velocities within the disturbance were not much greater than the minimum in tropical cyclones of the region, the storm's precipitation extended over an unusually large area.

The storm began moving inland near Panama City, Florida, at 8:00 P.M. on 19 June (see Figure 1.2). Its sustained velocities of 64–72 kilometers per hour in the coastal strip were sufficiently high to send watercraft scurrying for cover, to rip at loose structures, and to cause a storm surge that reached 1–2 meters above normal in some Florida west coast estuaries, although Agnes was by then no longer of hurricane dimensions. Fifteen tornadoes were generated along Agnes's borders on the Florida peninsula. As the storm moved northward over the coastal and Appalachian regions, it dumped heavy falls of rain. Along the rivers of Virginia, Pennsylvania, New York, and the New England states, warnings were issued for floods. Many basins had flows exceeding previous records. On the morning of 22 June the storm center was off the coast of Delaware and moving north. That night an unprecedented event changed the whole situation.

Agnes advanced northward across New York City and joined

TROPICAL STORM AGNES, 1972

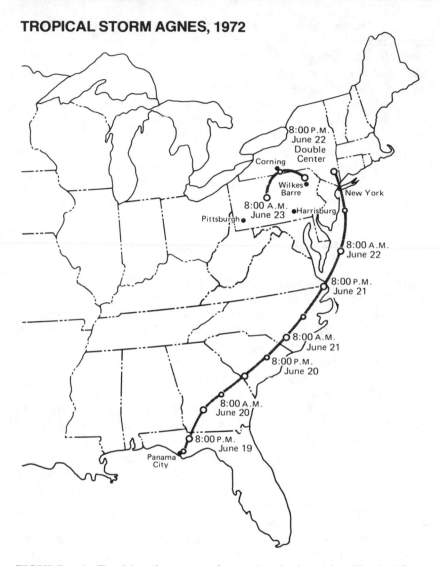

FIGURE 1.2. Tracking the storm—forecasting its intensity. Tropical Storm Agnes was observed by one of the most sophisticated systems of observation and forecast in the world, employing computer models, radar reports, satellite, aircraft, and radiosonde observations. Nevertheless, the turn westward on 22 June and the conjunction with a second weather system over Pennsylvania and New York was missed by the forecast model. The intensity of rain was not accurately predicted, and the combination of mainstream river flooding and flashflooding caused great confusion in the public understanding of such warnings as were given.

with a large system of colder air that had edged to the south and east. The result was a tremendous conjunction of rain-producing mechanisms moving westward across New York State and Pennsylvania. In some places, 19 inches fell over a 2-day period. Some forecasts of flood stage were accurate, but others proved far too low. A third of the stream-gauging stations in the Susquehanna Basin were washed out, as were the lines used to report their readings. At Harrisburg, Pennsylvania, river forecasters worked desperately by lantern light with a crippled communication system. At Pittsburgh the forecast office was moved to higher ground. Within 6 hours of beginning a new peak, downtown Harrisburg was under 3 feet of water from a tributary flash flood, the new Governor's Mansion was flooded above its first floor, and at many places in the region the streams were running higher than ever before recorded. At Richmond, Virginia, the James River was 2 meters above the record set in 1771. At Chemung, New York, the river of that name was 2.5 meters above the previous high of 1946.

By the evening of 25 June the waters were receding everywhere except on a reach of the Ohio. Agnes had spent her major force and had moved toward southern Ontario. The northeastern states were digging their way out of the destruction. Property damage was estimated to exceed $3.5 billion. At least 118 people had lost their lives. In monetary terms it was the greatest material disaster in the history of the United States.

At Wilkes-Barre in the Wyoming Valley of Pennsylvania, the levees erected by the U.S. Army Corps of Engineers to protect against floods of the magnitude of the previous record flood of 1936 had been surpassed. Following a warning the morning of 23 June, about 100,000 people were evacuated from the city before the flood walls began to cave late in the day. The densely settled Chemung Valley at Corning, New York, was likewise overrun by waters. Levees were destroyed at Richmond and Wellsville, New York. Approximately 40 percent of all the damage occurred in towns and cities where flows exceeded the capacity of protection works. Flows were far above the design levels, and the planned supplemental protection from upstream reservoirs had never been constructed. A huge extension of the detention storage system would have been required to contain the 1972 flood. In some municipalities such as Wilkes-Barre, a community disaster plan was used; other communities such as Corning had no plan; and in others the plan was not used or was obsolete. The whole region was disrupted by these unanticipated events. Power and transport were interrupted, farms were inundated, ancient graves and markers in Wilkes-Barre cemeteries were upturned, commercial districts were soaked with silt-laden waters. The Chesapeake Bay shellfish industry suffered as a result of reduction in salinity. More than 250,000 people in Pennsylvania alone were obliged to leave their homes.

The most serious damage was inflicted on a few industries and residents of the floodplains in areas such as the Wyoming Valley. Many of those industries had been located along streams to take advantage of conditions favorable for waterpower and waste disposal. For some manufacturers the need for streamside location had passed with the extension of municipal services; for others it continues to be essential. For dwellers in the floodplain the pattern of location was set in the early days of colonial settlement. Many present-day residents of Wilkes-Barre are retired coal miners or are otherwise connected with the declining anthracite industry. What was once an economic advantage—location on the floodplain—had become unnecessary for many of the flood victims, and yet thcy had little or no incentive or opportunity to move to another site beyond the reach of floods.

After having depended heavily upon flood protection from levees and reservoirs, the residents of the valleys deluged by Agnes were still trusting in the effective functioning of those works when that storm struck. By the end of 1972 they were looking vigorously at combinations of alternative measures, including warnings, insurance, and land-use management. Accuracy and speed of warning were seen as key elements in future action, and after Agnes steps wcre taken promptly to develop a genuinely integrated system to ensure that a suitable forecast could be disseminated to all users and that those users could be expected to act on the warning.

Although there had been a government flood-insurance program, including subsidized premiums for existing structures as well as actuarial rates for new buildings in hazard areas, before Agnes only a few hundred policies had been sold. Policyholders were indemnified for their losses; other sufferers had to rely upon personal savings, public relief, and government loans. A year later some of the uninsured joined with the Commonwealth of Pennsylvania in suing the Federal government for having failed to make the availability of insurance adequately known to the residents. This protest was especially significant because a condition for obtaining Federal insurance was preparation of plans and enactment of regulations by local authorities to provide curbs against further encroachment of flood-vulnerable construction in lower parts of the floodplain.

Thus, after Agnes the people and governments most affected

were now promoting insurance and land-use planning as a supplement to protection works. The traditional approach — meeting floods with flood control — had shifted to lively exploration of other tools, including protection works, warning systems, insurance, and land-use management. Responsibility for the disaster was spread among a large number of individual and group decisions that had been made over a long period. Yet much bitterness remained because, in a rich and technologically sophisticated country, warnings were inadequate, protection failed, and relief and reconstruction appeared agonizingly slow.

In the years following the Bangladesh disaster of 1970 and the Agnes disaster of 1972, the global scene changed in some respects and not in others. We review those trends and briefly examine how the situation has changed in those regions.

TRENDS IN LOSSES

Since we made our first attempts in 1978 to delineate the broad outlines of the world situation, natural hazard enlarged both in consequence and concept. The property and social organization vulnerable to the impacts of extreme events expanded in complex ways. Lamentably, accurate data on losses continued to be spotty and unreliable. The concepts of environment and hazard and the risks attached to them enlarged, while studies of the phenomena and their implications, described in Chapter 9, grew in precision and coverage. National and international action to cope with those enlarged risks strengthened in a few sectors but remained far short of what is required to avoid severe and lasting impairment of environmental quality and capacity for sustainable development.

Notwithstanding frequent appeals for improvement in the accuracy and coverage of estimates of losses suffered from extreme natural events, the quality of these data has not changed as rapidly as the number of organizations and programs directed at coping with the widely rising toll of losses. This is true of most countries, including the United States.

Much of the collection of data is based on poorly defined concepts of economic effects. Measurement of disaster is itself highly subjective (Horlick-Jones and Peters, 1991). For example, a published statement that an earthquake or a coastal hurricane "has caused property losses of X dollars" may differ from the true figure by a factor of two or three, and for a given area or type of event may not be comparable with an estimate for other areas or events. In our

experience, such popular estimates show a bias toward overesti-
mating losses from industrialized countries and underestimating
losses in developing countries or in areas remote from centers of
government and mass media. Recently, consistent, rigorous criteria
for estimating losses were suggested (Howe et al., 1991) but are not
employed in practice. Only when consistent criteria for assessing all
social losses—life and health, property, income, and environmental—are in place will there be a sound base for broad generalizations (see Green and Penning-Rowsell, 1988; Green et al., 1988;
Munroe and Ballard, 1983).

In view of these obstacles, we present the current picture of
global and national trends in four very rough and necessarily
incomplete ways. First, the record of major natural disasters
reported in the international press and government documents from
1900 to 1990 is reviewed. Second, more detailed estimates for
1947–1981 of area and fatalities by country are reported. Third, a
few estimates of global property loss are quoted. Fourth, in a series
of vignettes, we present estimates of loss of lives and property for
each of eight natural hazards in the United States. The latter are not
fully comparable but will serve to sketch the types of changes
underway. For each set we suggest major social factors that appear
to have some causal link with the apparent trend. With the recent
reappearance of major tropical cyclones on the East Coast of the
United States (Hurricane Hugo) and in the Bay of Bengal we use
this coincidence to ask how those regions, which were first examined
in the early 1970s, fared two decades later.

GLOBAL EVENTS

One way to gauge the shifts in the global experience with natural
hazards is to examine somewhat arbitrarily the total estimated loss
of life and the frequency of disasters causing more than 100 deaths.
We choose events reported to cause more than 100 deaths because
consequences of that magnitude are likely to be reported with
greater accuracy than lesser disasters. Those events for all countries
outside the United States have been estimated for 1900–1990 for
the Office of U.S. Foreign Disaster Assistance (1991) (see also
Heyman, Davis, and Krumpe, 1991). The figures surely are more
comprehensive and accurate for the past five decades than for the
four preceding decades and by 1992 are better suited to comparing
experience among different regions.

The greater number of deaths, as shown in Figure 1.3, occurred

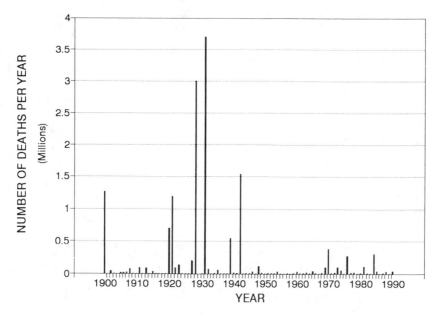

FIGURE 1.3. **Estimated deaths from natural disasters exceeding 100 deaths, 1900–1990.** Data drawn from press and government reports for all countries. Deaths from lesser events are not counted (U.S. FDA, 1991, and other sources).

in five major events. Disasters causing more than 500,000 deaths were drought in India (1900), drought in the Soviet Union (1921), and floods in China (1928, 1931, and 1939). These data exclude deaths from epidemics and war-caused famine that took heavy tolls. After World War II the principal causes of death were the Bengal cyclone (1970), the Tanshan earthquake (1976), and droughts in Mozambique (1981) and Ethiopia (1981).

When events causing more than 100 deaths after 1946 (as shown in Figure 1.4) are examined, we see that the number per year fluctuated between 3 and 28. The larger reported incidence in recent decades reflects in part the increasing coverage by reporting agencies. The data also suggest that as the world population increased, it became more vulnerable to natural disasters. During the same period the capacity of public agencies to cope with disasters enlarged sufficiently to prevent a repetition of the massive catastrophes that occurred during the earlier decades.

A more detailed analysis of disasters between 1947 and 1981 examined the occurrence of disasters, other than drought, over large areas, defined as having significant effects in areas of the average order of 10°-square grids (Sheehan and Hewitt, 1969; Dworkin,

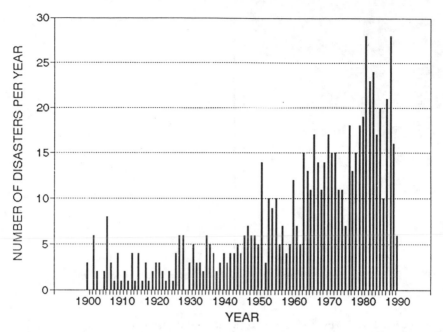

FIGURE 1.4. Estimated number of natural disasters exceeding 100 deaths, 1900–1990. Data for all countries are probably much more complete for the decades after World War II than for preceding years (U.S. FDA, 1991, and other sources).

1974; Thompson, 1982). Large-area disasters increased slowly from the late 1940s until the 1970s, decreased sharply, and then leveled off (see Figure 1.5).

Approximately 90 percent of the world's natural disasters causing more than 100 deaths originate in four hazard types: floods (40%), tropical cyclones (20%), earthquakes (15%), and drought (15%). Earthquakes probably are overestimated because they are easy to detect. Tropical cyclones cause somewhat less loss of life than earthquakes. Floods are most frequent and do the largest proportion of property damage. Droughts are very difficult to measure in extent, property damage, and death toll.

For the 1947–1981 period, excluding drought and excluding the Soviet Union, approximately 85 percent of all lives were lost in Asia, 4 percent in the Caribbean and Central America, and 4 percent in South America. The total toll was much smaller in Europe (2%), Africa (2%), North America (1%), and Australia (0.3%). (Drought is included in the data for Figures 1.3 and 1.4.)

In terms of proportion of population affected, the fatalities exceeded 1,000 per million in ten countries: Bangladesh (3,910), Guatemala (3,200), Nicaragua (2,600), Honduras (2,000), Indo-

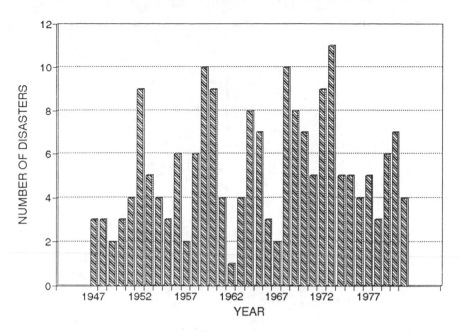

FIGURE 1.5. Large area disasters, 1947–1981. A five-year moving average of destructive extreme events, other than droughts, covering areas at least 10°-square grids.

nesia (1,500), New Guinea (1,300), Peru (1,300), Iran (1,500), Haiti (1,200), and South Korea (1,000). At the other extreme, twenty countries had fewer than 20 per million: Canada, Costa Rica, Bolivia, Uruguay, Belgium, Bulgaria, Czechoslovakia, Denmark, France, Hungary, Poland, Switzerland, Chad, Kenya, Nigeria, South Africa, Malaysia, Iraq, Israel, and Australia.

It is revealing to note how countries changed between 1973 and 1981 in their standing according to deaths per million and national income per capita (see Figure 1.6). In general, as GNP per capita increased, the number of fatalities per capita remained fairly constant for most countries. A few countries are charted on Figure 1.6 to illustrate the more common trends and deviant patterns, using logarithmic scales. In the low deaths–high income quadrant, Denmark, the Netherlands, and West Germany, like most others, made major income gains without changing mortality while the United States, Rumania, Greece, Indonesia, and Australia were among the few that increased noticeably in deaths per million. Most countries in high death categories held the same or lower approximate death rates as income per capita mounted; Japan and South Korea were examples. Some exceptions were Guatemala, Honduras, Peru, and Iran, where deaths also increased markedly with

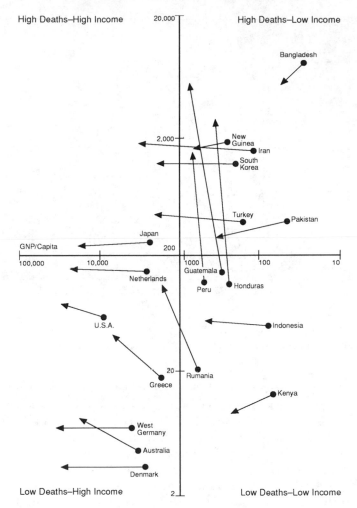

FIGURE 1.6. Deaths per million by national income. Changes in selected countries in per capita deaths from natural hazards and in per capita income, 1973–1986. The countries are selected to illustrate typical patterns of shifts, ranging from sharp increases in death toll to modest decreases (Thompson, 1982). (Income is expressed in GNP/capita [U.S. dollars] for 1981. Population is also based on 1981 figures.)

income. And declining deaths with increasing income were recorded in New Guinea, Bangladesh, Pakistan, and Kenya.

Far less reliable than estimates of loss of life are estimates of property damage. Calculating the costs of emergency action, of replacement of lost property, and of repairs of damaged property is a very difficult exercise and rarely uniform from one place to another. News media reports are notoriously inaccurate.

The data collected by insurance companies can be precise for claims made by damaged parties, but also are subject to great differences according to coverage and methods of estimating losses. The estimates by the Munich Reinsurance Company for the years 1960–1990, reproduced in Figure 1.7, show how the resulting losses may differ according to the cause of the disaster and from year to year. There is scattered evidence that the average volume of losses from all causes is mounting, that the fluctuations from year to year can be erratic, and that the proportion of losses compensated by insurance is increasing.

If accurate estimates for injuries, health effects, and property loss were available, the global picture might be different. Taking into account questions as to the validity and coverage of the data, it appears conservative to conclude that the numbers of hazardous events have increased, and that they reap a continuing harvest of deaths, but at a magnitude far below those in the first half of the century.

Some of the factors contributing to continued loss of life but reduction in the total numbers are the rapid growth of urban population in developing countries and the persistent invasion of

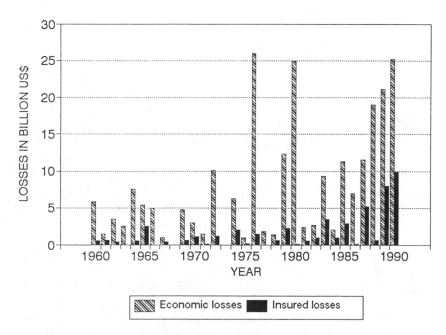

FIGURE 1.7. Economic and insured losses in 1990 dollars (major natural disasters, 1960–1990). The averages by decade for both losses and coverage mounted consistently. Drought is excluded.

coastal areas by agricultural, residential, and industrial activities. Although forecasting, warning, and social services improved in many areas, the numbers of people vulnerable to some type of extreme event grew. Technology had a dual role. It enabled more people to occupy hazardous areas. It also enhanced society's ability to respond or mitigate the consequences. There was great variation from year to year, and occurrences that might have gone unnoticed at the turn of the century were causing significant distress.

SELECTED HAZARDS IN THE UNITED STATES

The situation with regard to major natural hazards in the United States differs greatly among those hazards. The degree to which they affect the welfare of individual people or the national economy or the sustainability of the environment is influenced by natural conditions, social structure and process, and available technology. These occur in distinctive combinations from place to place (Showalter, Riebsame, and Myers, 1993).

Earthquakes, for example, occurred in severe magnitude in only a few places in the United States during 1960–1990. Following the Alaskan disaster in 1964 there was a shake at San Fernando in 1971 that caused loss of life through the failure of structures, and another at Whittier Narrows in 1987. Two years later in Northern California the Loma Prieta earthquake brought an estimated property loss exceeding $6 billion, and caused 62 deaths and at least 3,000 injuries. By comparison, the Alaskan disaster caused less damage but 100 deaths. Loma Prieta, combined with a false prediction of a major calamity along the New Madrid Seismic zone in Missouri (Showalter et al., 1991), stimulated broader public concern with efforts to mitigate the consequences of future disturbances. While interest in that direction heightened and more vigorous steps were taken in the Los Angeles, San Francisco, and Memphis–St. Louis areas, the pace of urban growth in those and other vulnerable areas continued to be rapid.

After relatively few landfalls along the Atlantic and Gulf coasts since 1969–1972, a series of severe tropical cyclones (hurricanes) struck in the 1980s (Figure 1.8). The most disastrous to that time — Hugo in 1989 — caused estimated losses in excess of $5 billion and 20 deaths in the Charleston, South Carolina, area. In that episode, as in other recent ones, the capacity to predict storms and storm surge proved relatively effective, as did the operation of evacuation and relief and rehabilitation services. Lessons learned from earlier

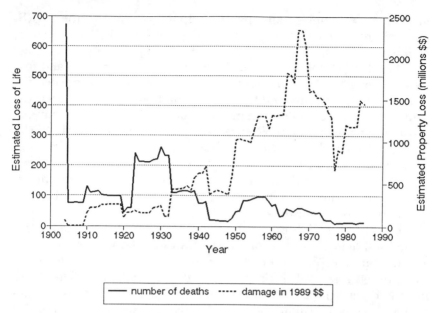

FIGURE 1.8. U.S. Tropical cyclone losses (10-year moving average, 1900–1990). (U.S. Weather Service, 1990).

disasters had been practiced, but new problems arose, as in the case of inland wind damage. As with earthquakes, the continuing occupation of vulnerable areas and the slow adoption of mitigation measures promised that a repetition of previous physical events would be likely to bring more damage and less loss of life. This was confirmed in 1992 when Hurricane Andrew caused property losses in the area south of Miami far in excess of Hugo. Deaths were estimated at 25 to 50 people.

Flood property losses also continued to rise during the 1970s and 1980s, after declining in the decade 1956–1965. The annual death toll ranged between 80 and 250, after reaching a maximum of 555 in 1972 (Riebsame, Diaz, Moses, and Price, 1986). In the more humid regions of the country, the wide deployment of floodplain management measures was offset by the continued vulnerability of older buildings, roads, and bridges and by further invasion of fringes of demarcated 100-year floodplains and small tributary floodplains in expanding urban areas. There, the problems of stormwater disposal, flood losses, and water quality became increasingly intertwined. In the semi-arid and arid regions of the country, those issues were made more acute by the difficulty of delineating shifting floodplains, by the very low frequency of intense storms, by increased peak flows, and by the rapid growth of cities (U.S. Federal Interagency Floodplain Management Task Force, 1992a, 1992b).

The 1987–1989 drought in the western United States, and to a lesser degree the 1985 drought in New York, was responsible for large loss to agriculture and for recognition that in some respects the country was becoming somewhat more vulnerable to extreme shortage of precipitation (Riebsame, Changnon, and Karl, 1991). This was believed to be due in various degrees to aging infrastructure, outmoded water management practices, and failure to keep pace with growing urban demands, as well as to inefficient monitoring and warning systems (Grigg and Vlachos, 1990). The crop losses in 1988 were estimated at $16 billion.

Deaths from lightning in the range of 60–100 annually continued at that level. Yearly tornado fatalities ran between 15 and 122, and the fatalities per event were lower in the decade of the 1980s.

Thus, in general in the United States, deaths associated with increased residential and recreational use of rural and open areas remained constant (lightning, flash floods, tornadoes), the increased use of such areas compensated by improvements in preventative actions. On the other hand, records for comparable disasters in large floods, hurricanes, and earthquakes show declining deaths and dramatic increases in property loss.

POLICY AND ORGANIZATION

Partly responsive to new research findings and partly reactive to public recognition that extreme events were causing hardship and inefficiency in ways that might be remedied, an impressive number of changes were made after 1974 in government policies and institutions and in research organization. The range of those changes is illustrated by the U. S. experience. Only a few other countries—notably Australia and Japan—have adopted as many different types of measures.

The capacity of the U. S. government to cope with earthquake, flood, hurricane, and tornado events was greatly enhanced by new legislation and programs. The more embracing and general measures were the Coastal Zone Management Act (CZMA) and the Federal Disaster Relief Act of 1974 (FDRA). The CZMA was the most ambitious effort to date to integrate the complex interests of federal and state agencies in preserving and using coastal resources of water, soil, biota, and scenery. The FDRA later provided through the Federal Emergency Management Agency (FEMA) for a more nearly unified system of planning for the delivery of emergency, relief, and rehabilitation services, including training and information.

In the earthquake field, the most influential policy was enacted in the National Earthquake Hazard Reduction Act of 1987 (NEHRA). This gave financial support, administrative coherence, and a rough integration to the work of agencies who were required to report their activities annually. It was re-enacted in 1991.

In the flood field, principal changes were undertaken in the new mapping and regulations under the National Flood Insurance Program. These strengthened requirements of positive action at the local level through community assessments and were supplemented by two executive orders—11296 and 11988—that provided that federal agencies would participate more in mitigation. A comprehensive assessment of the national experience with floodplain management was completed (U. S. Federal Interagency Floodplain Management Task Force, 1992a, 1992b).

In the hurricane field, an Interagency Hurricane Evacuation Program was put into effect and had critical testing in hurricanes Gilbert and Hugo.

A Federal Dam Safety Program undertook to require all affected federal and state agencies to identify and take remedial steps for any water storage structures in danger of failure with accompanying downstream losses.

An array of federal programs attacked the interrelated needs for information and education to promote understanding by officials, as well as public groups, of hazards they faced and appropriate action to cope. These included: FEMA preparation of public information materials; the FEMA/Building Seismic Safety Council earthquake hazard reduction series; the U. S. Geological Survey/ FEMA earthquake-awareness workshops; the National Institute of Mental Health's training workshops and materials for disaster workers; and the National Weather Service's hurricane, severe storm, and tornado education materials.

Much of this activity involved collaboration with state organizations and nationwide associations of local and state officials. Thus, on the national level the Association of State Floodplain Managers became a key means of promoting cooperation on flood problems, and the Association of State Wetlands Managers became the rallying group for a large group of hydrologists, engineers, biologists, and local administrators who addressed issues of wetlands definition, protection, and management. The Earthquake Engineering Research Institute similarly played a key role in mitigation efforts by agencies concerned with earthquake hazards. The Association of City Managers and the Council of State Governments took on new educational responsibilities.

Pioneering efforts were launched to cultivate cooperation among local, state, and federal agencies in working on regional situations. These included the Bay Area Regional Earthquake Preparedness Project, the Southern California Earthquake Preparedness Project, and the Central United States Earthquake Consortium. States and localities varied from giving systematic attention to the whole range of hazards, as in the case of Colorado, Tennessee, and Utah, to states which had little interest.

Research thrusts in some of the federal agencies were strengthened. In the USGS, for example, an experiment in Parkfield, California, tested modes of earthquake prediction and warning (Mileti, Farhar, and Fitzpatrick, 1990a; Mileti, Fitzpatrick, and Farhar, 1990b). The National Weather Service developed plans for improved warning services. These were linked with NOAA's extension of Doppler radar observation of short-term phenomena. The National Institute of Mental Health sponsored a synthesis of research on physical and mental health effects of disasters as well as encouraging new research (Ahearn and Cohen, 1984).

Whereas in 1970 there had been only three or four university centers that focused on natural hazards research, by 1992 the number in the United States had greatly expanded. The new centers included the Hazards Policy Program, Arizona State University; the Hazards Assessment Laboratory, Colorado State University; the Cornell University Natural Disaster Project; the Disaster Research Center, University of Delaware; the Center for Technology, Environment and Development, Clark University; the Hazards Research Center at the University of Louisville; the Center for Disaster Medicine at the University of New Mexico; the Hazards Reduction and Recovery Center, Texas A&M University; the Institute for Disaster Research, Texas Tech University; and the National Center for Earthquake Engineering at the State University of New York, Buffalo.

In other countries, the new institutions included the Asian Disaster Preparedness Center, Bangkok, Thailand; Australian Counter Disaster College, Victoria; Centre for Disaster Management at the University of New England, Armidale, Australia; Centre for Research on Epidemiology of Disasters at the Catholic University of Louvain, Belgium; Disaster Preparedness Resource Centre, University of British Columbia, Vancouver, Canada;

International Seismological Centre, Newbury, United Kingdom; Centro Nacional de Provencion de Desastres, Mexico D.F.; Middlesex Polytechnic Flood Hazard Research Centre, Enfield, United Kingdom; National Research Center for Disaster Prevention, Tsukuba City, Japan; Pan African Centre for Emergency Preparedness, Addis Ababa, Ethiopia; Pan Caribbean Disaster Preparedness and Prevention Project, Antigua, West Indies.

These national and regional efforts were supported with increasing strength by members of United Nations agencies. The United Nations Disaster Relief Office in Geneva facilitated international response to emergencies and collaborated with the International Committee for the Red Cross. Emergency preparedness was supported by activities of the United Nations Environment Programme, the United Nations Educational, Scientific, and Cultural Organization, the United Nations Development Programme, the World Health Organization, and the World Meteorological Organization. The Pan American Health Organization directs a special integration of disaster mitigation in its region. Further coordination of these and related activities was promoted by creation of a new office of Assistant Secretary-General.

Communication among research workers and between researchers and administrators responsible for using the findings was further enhanced by the establishment of the Natural Hazards Research and Information Applications Center at the University of Colorado, Boulder, and by the launching of publications giving attention to natural hazards alone or as a sector of the burgeoning field of risk analysis. These included, among others, *The Natural Hazards Observer*, *Risk Abstracts*, *The International Journal of Risk Analysis*, *The International Journal of Mass Emergencies and Disaster*, *Disaster Management*, *Mass Emergencies*, *Disasters*, and *The Journal of the International Society for the Prevention and Mitigation of Natural Hazards*, as well as publications in fields such as avalanches and earthquakes.

BANGLADESH AND TROPICAL STORM AGNES REVISITED

It is revealing to look at the experience in Bangladesh in 1970 and the eastern United States in 1972 from the perspective of tropical storms that visited those areas again in 1985 and 1989 respectively (Badolato et al., 1990; Baker, 1990; Haque and Blair, 1992; Mittler, 1991; Rubin and Popkin, 1991; U.S. Department of Commerce, 1990; U.S. Department of Defense, 1990; U.S. Federal

Emergency Management Agency, 1989; United Nations Development Programme, 1989; Rahman, 1991; U. S. General Accounting Office, 1990; United Nations Disaster Relief Office, 1991). From the appraising reports a large number of lessons are drawn in the interest of reducing losses in future (White, 1988). In both areas important improvements in forecasting, warning, evacuation, and relief procedures had been made following the earlier disasters so that losses of life were reduced below what otherwise probably would have happened. The elements of the problems faced remained, however, essentially as described above.

Bangladesh was visited by major floods in 1987 and 1988, the latter leaving about 25 million homeless. The 1985 tropical storm affected the southern coast with a storm surge bringing death to about 10,000. In April 1991 a wide-reaching tidal wave was believed to leave a death toll of at least 138,000. Striking in the harvest season, it killed at least 1 million livestock and caused heavy damage to crops and soil as well as to roads, and embankments, and over 700,000 homes.

Principal changes in human response in contrast to 1970 were that warning services were somewhat more effective and that this, combined with improved relief activity, led to a lower death toll than might otherwise have been taken by the cyclone. Shelter capacity was still insufficient, and the means of promoting constructive action by the recipients of warnings needed systematic improvement.

In September 1989, Hurricane Hugo visited a massive attack of wind and water on Puerto Rico, the Virgin Islands, and the Carolina coast, carrying severe storms inland. River flooding was not a major consequence. Although Agnes and Hugo had different physical configurations, certain of the problems presented by the two were similar. The Hugo forecasts of landfall were moderately accurate. Except for the Virgin Islands, where preparedness had been low, the immediate response by local and state governments was to provide fairly effectively for evacuation and emergency services such as food and shelter. Deaths in the continental United States were lower than in 1972. Staffing and coordination problems arose in the early days and hampered management of the recovery phase. The experience revealed needs to modify training to deal with recovery activities, to improve means of providing individual assistance, and to strengthen preparedness measures (U. S. General Accounting Office, 1991). Property losses were estimated to be larger than in previous storms and were most heavy in the coastal areas. Steps were taken to clarify roles of the several levels of government agencies, to strengthen federal organization and early response capacity, and to inform and train local officials who are involved.

FEATURES IN COMMON

These four storms, moving though they did across the coasts of nations with radically different cultures and social organization, exhibited six features in common. First, the exposure of large populations to the fury of the tropical cyclone was not the result of casual or ignorant activity. Vulnerability to the risk of a destructive storm was the corollary of seeking beneficial use of land resources. Increased hazard accompanied increased material wealth. Each situation represents a trade-off between economic return and social risk.

Second, the heightened hazard from the extreme events was the product of interaction of natural systems and human use systems. Bangladesh and the southeast United States are very different. Bangladesh strives through development activity to expand its crop lands. The United States has a high degree of industrialization — albeit declining in several of the affected areas. In each case, however, heavy reliance upon single technological tools in the form of substantial engineering works to cope with extreme events led to increased productivity as well as to unprecedented disaster.

Third, the choice of those courses of action was made by a combination of community and individual decisions within a framework of forces that neither individual nor community could control. The tragic episodes of the four cyclones were neither wholly the product of government fiat nor wholly the sum of many individual choices. The Bangladesh peasants, like the retired American coal miners living in the Wyoming Valley, exercised some initiative within the limits set by public policy and social opportunity.

Fourth, the disasters by their very nature were inequitable. The contrast between survivors mourning their dead, victims trying to re-establish home or farm, and fellow citizens modestly sharing in the loss of others is enormous, fundamental, and perhaps immeasurable. Inequity was not randomly distributed. It was not mere chance that the burden fell most heavily on the landless laborer of Bangladesh or the old retiree of Wilkes-Barre.

Fifth, the material damage, loss of life, and social dislocations exceeded what would have occurred had reasonable measures been taken in advance of the storms. Major disruption was inevitable whenever population was in the path of such forces, but it might have been reduced in significant ways by actions involving warning systems, building design, and land-use planning. Any of these measures, effectively used, could have reduced losses at relatively small cost.

Sixth, each disaster generated measures to prevent its repetition. In Bangladesh the warning system was improved. In the northeastern United States, federal legislation was enacted to encourage land-use planning and insurance in the floodplains.

These features are duplicated in numerous hazard situations elsewhere. The Bangladesh cyclone and Tropical Storm Agnes are not unique in that sense, but they exemplify dramatically conditions prevailing in different forms in other countries with different types of development and with other hazards. It is usual to find that the disaster area had been occupied to gain certain benefits of location, that natural and social systems interact to produce a hazard, that the choice had been made by both individual people and public agencies, that the losses might have been reduced by inexpensive preventive action, and that changes in policy promptly follow.

For all these features in common, there is one very important difference. In the United States 250,000 people were evacuated in 1972 and only a dozen died. In Bangladesh in 1970 at least 225,000 lost their lives. To be poor as a nation or a person is to be particularly vulnerable.

The rising toll of damage—whether in Bangladesh or in the United States or in any other nation—may be explained largely by three forces: (1) the spread of people; (2) the increase in recorded disaster events; and (3) the enlarged hazard of developing countries.

The Spread of People

During the 90 years after 1900 the earth's population increased by 4.8 billion. Moreover, by extending its presence in more places and in larger numbers than before, the human race exposes itself and its artifacts more widely to risk from natural events. In addition to the gain in population, however, two other trends at work are almost everywhere. First of all, with accumulation of material wealth mounting in most countries, there is simply more property to be damaged. Correcting for changes in monetary value, the net worth of private households in the world is estimated to have risen from U. S. $404 billion in 1882 to $1,256 billion in 1952—a threefold increase (Doane, 1957). Despite the weakness of such statistics, there is indicated an average annual increase in total wealth of about

3 percent, which would bring the 1975 figure to about $2,479 billion.

The second trend influencing social susceptibility to natural hazards is spatial distribution of population. In virtually all countries the predominant movement of the past half-century has been from farm to town or city. In some areas, such as in the outmovement of peasants from Brazil's drought "polygon" to the favellas of southern cities, this has reduced the number of people at risk at the same time that total rural population increased. It also concentrated people in urban areas highly vulnerable to other hazards. For example, in Tokyo a combined earthquake, fire, and tsunami (tidal wave) may one day take a heavier toll of lives than if the migrants had remained in their native rural communities. In industrial countries the outward movement of suburban population, coupled with automobile transport and the building of second homes, may bring city folk to habitats with which they are unfamiliar. Austrian city dwellers, for example, are building country homes in the path of avalanches that their country neighbors know enough to avoid. Americans in search of sunny beaches are locating on Caribbean shores subject to tropical storms that most of them have never experienced.

Detailed estimates of what has happened in selected countries are presented in Chapter 3 (for the world scene the data are so incomplete that totals must be taken as a rough approximation). The studies in these countries and local areas support our judgment that property losses generally are rising. A more difficult problem is estimating the extent to which such loss and the cost of adjustments are offset by gain in material goods, income, and social stability.

More Disasters, Fewer Deaths, Potential Catastrophes

Two concurrent trends illustrated in Figures 1.3 and 1.4 appear contradictory. There are many more disasters where deaths exceed 100, yet the total number of deaths is lower. The increased frequency of disasters, as noted above, is probably due both to improved reporting in recent times and to the increase in population and its concentration in hazardous areas. The decrease in catastrophic death tolls (over 500,000) is probably due to improvements in warning, evacuation, and disaster prevention. Still, the potential for catastrophe remains. Pressure for living space, the spread of cities, confidence in the role of technology in protecting people from the natural environment, and lack of personnel skilled in the planning arts and in dealing with the extreme possibilities of environment all point to the liklihood of future catastrophe (see Figure 1.9).

Expanding Hazard in Developing Countries

Modern tropical cities spread onto floodplain lands just as their counterparts have done in temperate latitudes. Thus, in Sri Lanka much of the growth of Colombo during recent years was residential and industrial construction in the floodplain of the Kelani River. Again, in Kingston, Jamaica, just as in the San Francisco Bay Area, new buildings are constructed on alluvial and unconsolidated soils, highly unstable in earthquakes. In Tanzania, the rising demand for

Tokyo Earthquake

The Great Kanto Earthquake of September 1, 1923, started a fire that burned for three days in Tokyo, killing more than 59,000 people and destroying some 370,000 buildings. Fifty years later, Tokyo was the world's most populous city, with five times as many people as in 1923, full of facilities where gasoline, oil, liquified gases, and chemicals are stored, and with 80% of the city still crowded with wooden houses. Some 65 km² of the city are below sea level and subject to flooding. A recurrence of the 1923 earthquake could result in hundreds of thousands of deaths in the absence of preventive measures that are only now being undertaken.

Danube Flood

In the winter of 1838 an accumulation of ice in the Danube Valley combined with high flows to destroy nearly half the buildings of Budapest. Since then the Hungarian National Water Authority has constructed Europe's largest system of levees and dikes, and has organized a flood-fighting service. Experience with a great flood along the Danube in 1965 and in the valley of a tributary, the Tisza, in 1970 (100,000 people were evacuated and 30,000 were engaged in flood fighting) suggests that greater floods will occur and are likely to overtop the present works. The danger is increased by a tendency for river levels to rise as a result of flood protection works and of land-use changes upstream, chiefly in neighboring countries.

Afro-Asian Drought

In 1972 lack of rain in many countries led to poor crops. World grain production fell by 4%. Drought in the Soviet Union caused that country to buy one-fourth of the U.S. wheat crop in 1973. World reserves of grain have dropped from a 95-day supply in 1961 to a 26-day supply in 1974. At the same time, some climatologists fear a continuing period of unfavorable weather. Thus, simultaneous flood or drought in either the major grain producer nations or wealthy consumer nations coupled with a drought in the monsoonal areas of Africa and South Asia, could create famine conditions on a scale greater than ever recorded and one that would exceed the world's capacity in surplus grains.

FIGURE 1.9. Future catastrophes.

urban foodstuffs and the heightened dependence upon central marketing make the central government more responsible for providing drought relief. In general, the more successful the development process in other respects, the higher is the loss from natural hazards likely to rise. If it continues to parallel the "western model," this economic process will do much to help multiply disaster situations in future.

Development aid from the world's richer nations may exacerbate inadvertently this rising toll. The partial protective works in the deltaic lowlands of Bangladesh, responsible in part for cyclone-vulnerable settlement, were financed by the Netherlands and the United States. The centralized grain marketing of Tanzania, which increases drought-cost potential, were further centralized by construction of "modern" silo and storage facilities imported from Sweden. Excessive reliance on dikes (with subsequent failure) in Sri Lanka was learned from English engineering books by students on United Kingdom fellowships.

At the same time there is little sign that modernization can relieve traditional societies of their high death toll — at least in the short run. While the growth of cities is alone sufficient to generate rising flood loss, the provision of communication/information services and transportation facilities in rural areas is not sufficient to reduce loss of life from major disasters to the degree that this has been achieved in the industrial nations. When a hurricane strikes the Gulf Coast of Louisiana or Texas, many thousands of people can be safely evacuated in a few hours. Despite accurate forecasts by satellites, the years 1970 and 1991 demonstrated that it is still difficult to achieve comparable evacuation in Bangladesh and in similar cyclone situations in developing countries. The changes already under way in developing countries seem likely, if permitted to continue, to bring increase in economic loss from natural hazards in the modern sectors of the economy while failing to offset high loss of life in the more traditional sectors.

Yet natural-hazard problems should not be considered of low priority in developing countries (Kates, 1980; Cuny, 1983). A major drought can sometimes diminish crop yields in Tanzania by as much as 30 percent, which is equivalent in gross national product terms to a 4 percent reduction. Reduction of half this loss would mean a return equivalent to 20 percent of the total current investment in development. Similarly, in Sri Lanka, where, in an average year, floods cause an estimated reduction of $30 million in the value of agricultural crops for domestic consumption, the elimination of 50 percent of all flood damage would be equivalent to raising the national growth rate by 0.60 percent (Hewapathirane, 1977). These are significant figures in nations where economic gains are small and hard to achieve.

A government policy that would curb uneconomic exposure to extreme events and thereby reduce the loss from natural hazards in developing countries would be a worthwhile investment if it entails relatively low costs and especially when, in the absence of counter-acting policies, a rising toll from such hazards seems likely. Without such a policy, year-to-year losses will rise and the disruption from occasional disasters will become greater still. If no new action is taken now and if efforts are not made to develop an appropriate policy, the culpability for increasing damage tolls, added to an already heavy loss of life, will rest entirely with human acts of omission and not with the extreme acts of commission that occur in the natural environment.

With the mounting potential for catastrophe from individual events, the probability of a conjunction of events is also heightened. By way of illustration, a great drought in the Indian subcontinent, coinciding with heavy floods in China and an earthquake in Japan, would bring about much heavier damage than in the past. Each could heighten the consequences of the other by making demands upon available supplies and by curtailing production for world trade.

A CRUCIAL TIME

Most disasters have certain features in common: losses are rising, frequency is increasing, and the burden of loss and adjustment costs falls quite inequitably among the world's nations. Does this mean that the natural environment is becoming more hazardous? Our assessment, as reported in detail in Chapter 8, is that indeed the natural environment is becoming more hazardous in a number of complex ways that defy immediate or easy reversal of the process. To gain an appreciation of the conclusions and to understand the reasons that the process cannot now be readily reversed, we have organized supporting evidence in seven analytical steps, in the following chapters.

At the outset the concepts of extreme events and hazard require definition, and the patterns of human response and choice require description. Review of the ways in which people may cope with extreme events then leads to the realization that some kind of choice always is involved. The range of experience in actually coping with hazard provides one ground for asking how nations might respond in the future. This experience can be summarized by comparing the ways a high-income country and a low-income country have dealt with a particular hazard. Agricultural drought, floods, tropical cyclones, and air pollution are examined as representative of a variety of hazard situations. The consequences of hazard are seen to vary both by hazard and by income.

With the language of Chapter 2 and the experience of Chapter 3, each of the three succeeding chapters examines the process of human occupance for different perspectives of scale, beginning with the processes by which individuals and groups choose among the alternatives open to them. At the individual level it should be recognized that people are not alike in appraising a hazard, in perceiving their options, and in deciding what to do. At the level of collective action it is helpful to identify the ways in which communities, communes, and corporations guide or manage or serve the adjustments practiced by individuals. As with the range of experience, these methods are illustrated with practical examples from both high-income and low-income countries.

A nation may, of course, arrive at a particular mix of adjustments in coping with extreme events without consciously adopting a policy. There are at least five types of guides that may be pursued in fostering or preventing action by individuals and groups. These show signs of progressing in similar directions. Against this background, it is desirable to ask what is worth sharing among nations and what is worth doing jointly. Which types of adjustment are widely practicable and which are not? Which efforts can be effective only if based on international collaboration?

In reaching the point of suggesting what considerations should enter into formulation of national policy and international action, it is essential to recapitulate the theory within which the analysis unfolds. Previous exploration of hazard proneness emphasizes either nature, technology, or society. Our approach emphasizes interaction among all three. Like appraisals of national experience and of individual and collective choices, this theory is far from adequate. It is nonetheless a beginning at identifying different modes of coping with natural extremes as well as the major factors affecting the mix of adjustments at a given place and time.

With evidence from field observations and in the framework of provisional theory, we may now return to the opening question as to the hazardousness of the environment. The tentative answers suggest the prevailing trends and the conditions in which the prospect might be altered.

It seems possible for developing nations to avoid making the more extreme mistakes displayed so tragically by an industrialized society in preparing for Tropical Storm Agnes and Hurricane Hugo. Neither rich countries nor poor countries need endure a long period of increasing uneconomic property damage and catastrophe arising from environmental extremes. And yet, for reasons that will be unfolded, the globe's inhabitants are likely to suffer continuing loss with increasing vulnerability to catastrophe. It is yet to be seen whether, as people multiply their technical options, the hazard enlarges rather than diminishes.

Hazard, Response, and Choice

Whether or not a Bengal fisherman behaves like his neighbor in the face of the roaring cyclone, his action or inaction can usually be illuminated by examining three elements in the situation. These three are the ways in which people (1) recognize and describe a hazard, (2) consider how they might deal with it, and (3) choose among the actions that seem to them available.

In moving toward an understanding of hazard, response, and choice, we must define the hazard itself, and what is meant by response to hazard and by choice of response. These concepts of response and choice provide a basis for analyzing the experience with hazards in selected countries.

EXTREME EVENTS IN NATURE AND NATURAL HAZARDS

A basic distinction has to be made between extreme events in nature, which are not necessarily hazardous to people, and the character of hazard events. The natural events system—the array of wind, water, and earth processes—functions largely independently of human activities and is an object of scientific inquiry in its own right by meteorologists, hydrologists, and geologists. Conversely, for practical purposes large parts of the social system may also be regarded as operating independently of natural events. Interaction of the two creates resources. It also creates hazards or negative resources. This distinction is diagrammed in Figure 2.1.

An extreme event such as a lightning stroke or a flood may be a productive resource and a hazard at the same time. Lightning may kill an animal but also start a fire essential to the preservation of a forest ecosystem. A flood may destroy a farmstead while fertilizing the fields. The hazard is the risk encountered in occupying a place subject to lightning or flood. The actual hazard, not the natural

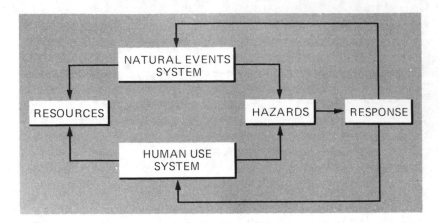

FIGURE 2.1. Resources and hazards from nature and man. The interaction of nature and man creates both useful resources and hazardous threats for human beings. Responding to those hazards, society may seek to modify the natural events system, the array of wind, water, and earth processes, and the human use system of locations, livelihoods, and social organization.

event, is the present subject of inquiry. Although the hazard results from the interaction of natural and social systems, the two cannot be equated as causes. Natural systems are neither benevolent nor maliciously motivated toward their members: they are neutral, in the sense that they neither prescribe nor set powerful constraints on what can be done with them. It is people who transform the environment into resources and hazards, by using natural features for economic, social, and aesthetic purposes.

There are many ways of describing extreme natural events. In several instances the study of these phenomena is a central concern of one or another branch of the natural sciences. Thus, hydrologists study floods; seismologists study earthquakes; droughts and tornadoes are the concern of climatologists and meteorologists; patterns of land and land use are studied by geographers; biologists investigate the desert locust; and so on. Out of this traditional division of labor comes a classification of hazards based on the natural processes characterized by extreme events. Such a classification, as presented in Table 2.1, has been used for the organization of research in the natural sciences. It invites description of natural events according to those characteristics considered appropriate for study by physical and biological scientists, but is less appropriate when applied to the question of social response.

When the same events are described as hazards, quite different characteristics may become important or paramount. Meteorolog-

TABLE 2.1. Extreme Events, by Principal Causal Agent

Geophysical		Biological	
Meteorological	Geomorphic	Floral	Faunal
Blizzard and	Avalanche — rock	Fungal disease	Bacterial, viral,
snow	Avalanche — snow	*(examples)*	and protozoal
Cold wave	Earthquake	Athlete's foot	disease
Drought	Erosion	Dutch elm	*(examples)*
Flood	(including	Wheat stem rust	Influenza
Fog	erosion, and	Blister rust	Malaria
Frost	shore and beach	Infestation	Typhus
Hailstorm	erosion)	*(examples)*	Bubonic plague
Heat wave	Expansive soil	Weeds	Venereal disease
Lightning	Landslide	Phreatophytes	Rabies
strike	Shifting sand	Water hyacinth	Hoof and mouth
and fire	Tsunami	Hay fever	disease
Temperature	Volcanic eruption	Poison ivy	Tobacco mosaic
inversion		Red tide	Infestation
Tornado			*(examples)*
Tropical cyclone			Rabbits
(hurricane,			Termites
typhoon)			Locusts
Windstorm			Grasshoppers
			Venomous animal
			bite

Source: Adapted from Burton and Kates (1964), p. 415.

ical analysis of heavy snowfalls commonly relies upon description of events in terms of depth of snow accumulation, the water equivalent of snow, and the accompanying conditions of wind and temperature. For the purpose of understanding human response to snowfall, however, a number of other physical characteristics are important (Rooney, 1969). These include the frequency with which severe, disrupting snowfall can be expected in a given community, the time of occurrence by hour of the day and by day of the week, and the period elapsing between onset and the peak fall (Archer, 1970; Baumann and Russell, 1971). The relation between snowfall characteristics and impact is not a simple linear function: it depends upon the ways in which the people of the area commonly cope with the event. Snowfall below a critical threshold value may not cause any significant damage or disruption. Once a critical threshold has been passed, however, damage may mount rapidly.

The specification of these relations and the definition of the threshold levels for a given place or society pose significant problems for research not normally approached by physical scientists. A threshold of crippling snowfall for Toronto, for example, is lower than the threshold for northern Ireland. Indeed, the common units

of measurement employed for physical delimitation may be un-
suited for assessment of social impact. Where the units are appropriate, an accurate measure of social significance of hazard may be
gained only by a specific *combination* of such measurements and
requires research on both physical and social systems.

ENVIRONMENTAL PARAMETERS
FOR HUMAN RESPONSE

The search for measurements of physical phenomena appropriate to
analysis of human response is carried out in the margins between
physical and social sciences. The characteristic that defines the event
is its *magnitude*: only occurrences exceeding some common level of
magnitude are extreme. Describing the magnitude of an event often
calls for imaginative use of data obtained for other purposes or, in
some cases, for generation of new types of measure. Other common
measures of hazardous events found to be significant in human
terms are *frequency, duration, areal extent, speed of onset, spatial dispersion,*
and *temporal spacing.* The first four of these are measures of the
aggregate of separate events. Spatial dispersion and temporal
spacing refer to the distribution of a population of events over space
and time.

The *magnitude* of a flood, for example, is the maximum height
reached by the flood waters or the maximum discharge at a given
point. Stream discharge is not considered a flood until the water
reaches a magnitude that is overbank (there are more sophisticated
measures for purposes of frequency calculations).

Frequency asserts how often an event of a given magnitude may
be expected to occur in the long-run average. Thus, a snowstorm of
a given magnitude (depth of snow accumulation in cm) may occur
only once in ten years (but not, of course, every ten years). To state
that a storm or flood or killing frost has a recurrence interval of 10
years is to say it has in any year a 10 percent chance of occurring.

Duration refers to length of time over which a hazard event
persists. Thus, an unexpected frost may cause much damage in a
few hours; a fog may persist for days; a flood may last for weeks;
and a drought, for years.

Areal extent refers to the space covered by a hazard event. A
tornado's areal extent may be a short and narrow swath, while a
drought may cover thousands of square kilometers.

Speed of onset refers to length of time between the first appearance of an event and its peak. Soil erosion and drought are

slow-onset hazards, whereas earthquakes and avalanches are fast-onset hazards.

Spatial dispersion refers to pattern of distribution over the space in which it can occur. Droughts and heat waves are usually much more widespread in occurrence than other events. Avalanche paths, eroding coasts, and floodplains ordinarily can be mapped with some precision. It is usually possible to identify areas prone to specific hazards, but there are difficulties, as illustrated by the question of whether seismic risk is large in an area where tectonic movement is possible but has not been recorded for two centuries.

Similarly with *temporal spacing* — the sequence of events. Some hazards (like volcanic eruptions) approximate a random time distribution, while others (e.g., tornadoes) are seasonal or cyclical, and in some areas severe earthquakes occur only after long quiescent periods.

The significance of the seven dimensions of hazardous events is to be measured in terms of the type of response each requires or allows. Magnitude, frequency, and areal extent describe the strength or force of an event, how often it can be expected to occur, and over what area. Generally speaking, the greater or more powerful the hazard event, the less adequate the available technology to control or mitigate it. The more frequently the hazard occurs, the greater the need to take steps to respond to or accommodate it. The larger the area affected, the broader the segment of society likely to be subjected to loss or disruption.

The significance of the speed of onset dimension is chiefly in terms of emergency preparations and of the physical capacity to operate a warning system. Where a hazard event strikes rapidly, little can be done; where a long period of time elapses between onset and peak, the range of possible responses is usually correspondingly greater. The ability to predict the event may be based on observation of precursory phenomena such as air masses or tectonic stress.

Much the same applies to duration. The shorter the event in time, the less can be done during its occurrence. Hazard events that last for days or weeks or longer, as in the case of the Murray River floods in Australia, allow for more mitigating actions to be taken while the hazard event continues, but also test the society's capacity for endurance and extended change in behavior.

Spatial dispersion is influential chiefly in determining over what territory a pattern of hazard-response is needed. Thus, adoption of measures to mitigate the effects of earthquake depends on knowledge of where earthquake is likely to occur. Temporal spacing carries implications especially for the scheduling of human activities. Where hazards are seasonal in occurrence, for example, it may be possible to concentrate hazard-vulnerable farming activities in the non-hazard season or to schedule preparations for the physical event; and the conditions in one period may influence what the farmer does in the next (Curry, 1962).

Each of these dimensions of hazard events presents complex problems of measurement, and each affects profoundly the kinds of human responses that are appropriate. All are important, but magnitude deserves to be examined first and in greatest detail because it delimits the essential characteristic of the extreme event.

The Measurement of Magnitude

For almost all natural events, especially at the extreme end of the spectrum, some measurement of magnitude exists. The magnitude of hazard events is often particularly difficult to assess, however, because the instrument of measurement—persons, their possessions and activities—is itself subject to significant and rapid change. Thus, two droughts may be of similar magnitude in terms of moisture deficiency, duration, and areal extent, and yet have very different magnitude in social response terms according to density of population, kinds of crops being grown in the area, level of agricultural technology, and so on. Both physical- and human-significance scales of magnitude exist for a few extreme events.

In Figure 2.2 the Richter and the Modified Mercalli scales of earthquake magnitude are set out for comparison. The Richter is a complex logarithmic scale expressing the amount of displacement of the seismograph pen and the distance from the earthquake's epicenter and indirectly reflecting the amount of energy released and the amplitude of the seismic waves. A whole unit is estimated to be 10 times the amplitude of the seismic wave of the previous value and many times greater in energy release. The largest earthquake recorded (8.9) may be 700,000 times greater in terms of energy released than an earthquake (Richter scale 5.0) that approximates lower levels of significant damage (Modified Mercalli scale VI).

The Modified Mercalli scale measures the intensity of an earthquake as experienced. Thus, its measure is not of the physical universe but of impact upon man and his works. The measure of intensity described by the Modified Mercalli scale is the ability to sense the earthquake without instruments, to observe its effects by ordinary people (a practiced observer is preferred), and to describe its impact on people and their possessions.

The magnitude scale (Richter) is often presented as a scientific

Intensity Scale

MODIFIED MERCALLI NUMBER	I	II	III	IV	V	VI	VII	VIII	IX	X	XI	XII
			Persons						Structures			
PERCEIVED BY:	None	Few	Some	Many	Most	All						
DAMAGE TO:					Glass Plaster	Furniture Chimneys	Poor	Ordinary	Resistant	Many	Most	All
DESTRUCTION TO:										Some	Many	Most

Magnitude Scale

RICHTER NUMBER	1-2	3	4	5	6	7	8
ENERGY RELEASE IN ERGS:	4.47×10^{12}	7.94×10^{14}	2.51×10^{16}	7.94×10^{17}	2.51×10^{19}	7.94×10^{20}	2.51×10^{22}
IN MULTIPLES OF BASE:	1-31.6	1,000	31,600	1,000,000	31,600,000	1,000,000,000	31,600,000,000

FIGURE 2.2. The measurement of an earthquake—comparing intensity and magnitude. Intensity is a measure of the human experience and impact of earthquakes; magnitude is an estimate of energy release. They are roughly comparable, as shown. With remote seismographs magnitude can be estimated for almost all earthquakes but, in the absence of people or their property, there is no meaningful measure of intensity.

scale, whereas the intensity scale (Modified Mercalli) is viewed as subjective and, by inference, as less reliable. In their own ways, however, both are scientific, and both are constrained by the human limits of science. Energy release is as arbitrary a measure of bigness as is human perception or impact. Even though the use of one parameter seems more appropriate for a particular purpose, the scales are subject to change to meet other requirements in design of structures. Japanese and American scientists scale earthquake magnitude in different ways. No less than eight different scales of intensity are in current use on the international scene. The magnitude estimates of the Great Alaska Earthquake of 1964 range from 8.3 to 8.75 on the Richter scale, a difference of some 4.7 times in energy release.

The ways in which earthquakes or other extreme events are measured is very much a matter of cultural convention, scientific knowledge, data availability, and functional need. For present purposes the ideal measure of earthquake bigness—the dimensions of the hazard rather than the event—is an estimate of the damage potential posed by the event. The ideal cannot be met by current seismological data, despite widespread usage of both physical and human scales and the relatively large effort expended by the world's seismological community on magnitude and intensity measurement.

For most other hazards, the issue of bigness is described in physical terms related to some social parameter. For example, the magnitude of the flood is commonly described in terms of peak discharge. Thus, it can be stated that in Rapid City, South Dakota, the maximum peak discharge of the flood of 9 June 1972 on Rapid Creek was 5,400 cubic meters per second (or 50,600 cfs), the discharge being calculated for a measured cross section with an observed water level. The tropical cyclone of November 1970 in Bangladesh had a barometric pressure at its center as low as 950–960 millibars, winds of up to 125 kilometers per hour, and a storm surge of 3–7 meters.

Statements of physical size can be related to damage potential by various transformations. The Rapid Creek river flow is expressed as inundation depth on the land for a designated area (Figure 2.3). Given depth and area inundated, and given also the human use of that area, it becomes possible to estimate the damage potential. In the case of cyclones there are more complicated relations, involving the size of sea surge, discharge of rivers, the offshore slopes, wind velocity, and direction of the coastline.

Although there is a rough relation between magnitude and frequency for events of lesser magnitude, when it comes to the

HYDROLOGY AND ECONOMICS
From Peak Flow to Damage Estimate

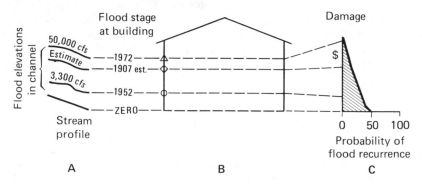

FIGURE 2.3. Estimating flood damage at Rapid City, South Dakota. Assuming that the record of past floods is a helpful, if incomplete, guide to the future, the analyst begins by drawing a profile of flood elevations along the stream channel (A). For each damageable structure and its contents, estimates of potential in damage are made (B). These in turn are related to the probability that a flood equal to or greater than recorded floods might occur again in the future. The probabilities are expressed in the percent likelihood that such floods might occur in any year. The area under the resulting curve of damage and probability yields an estimate of the average annual damages expected over the long term (C).

events of very large magnitude (and therefore extremely low frequency), it is difficult to assign any probability to the event. If, for example, interest lies in anticipating a flood such as that of the Arno at Florence in 1969, having a roughly computed probability of once in four centuries, there is great difficulty in checking the validity of the estimate, even with such long-kept records as those in Italy. This problem becomes even more acute in countries with a shorter length of record and in dealing with events of lower frequency. A major earthquake may well occur only once in a few thousand years in a given seismic locality, and ordinarily, outside of China the historical record is not long enough to check the estimate. Evidence from tree rings, clay varves, and radiocarbon dating may extend the record, but often it is difficult to assemble.

In these circumstances it may be advisable not to depend on stochastic methods but, rather, to attempt to identify causal relations on the basis of which it is possible to arrive at an estimate. A case in point is the work done by hydroelectric power planners on a stream in Venezuela for which there were records going back only a decade or so (Byers and Braham, 1959). The magnitude of the maximum flood for which provision should be made on the spillway of the dam was estimated without reference to historic records, but was based upon a computation of the maximum possible conjunction of atmospheric phenomena producing high-intensity rainfall over the area, and extrapolating from that the maximum possible run-off. A somewhat similar requirement applies for most other natural events in the extremely low frequencies. The estimate of causal relations may be far more significant than the extrapolation of probabilities.

Comparative Magnitudes of Hazards

So far no adequate scale has been devised to compare the magnitudes of different types of natural hazard. A hypothetical scale developed by Hewitt (Hewitt and Burton, 1971) seeks to convert, for specific places, the varying levels of some common physical measurements to "damage energy," a new measure of magnitude reflecting damage potential. The new scale remains largely untested.

In the absence of a measure of this kind, estimating energy release is one way of comparing hazards. Extreme events can be gauged in terms of energy release per unit of area or per unit of time, and either would serve as a surrogate measure for magnitude. Earthquake is an example of high-energy release per unit of area and of time. The energy released in a tropical cyclone is also very high but is spread over a wide area and a longer time. It is evident that a statement of the energy released per unit of time and of area for the Alaska earthquake of 1964—or for Hurricane Camille (1959)—has no necessary connection with the human impact of those two events. In terms of possible and appropriate human responses, the differences between such hazards in the degree of energy release are significant because of the magnitude of disturbance.

In theory it is possible to rank or group all hazard events according to their several characteristics. In one device for such classification (shown in Figure 2.4), for three events—a drought, a blizzard, and an earthquake—a profile is drawn according to its position on the six scales found to be most significant in terms of human response for a given society. Typically, a major earthquake is infrequent, of short duration, and relatively concentrated in extent. Speed of onset is usually very high; the earthquake zones of the earth are limited; and the event occurs in a more or less random fashion. Compared with a major earthquake, a severe drought event is of greater frequency, much longer duration, more widespread over the globe, slower in onset, more diffuse, and somewhat

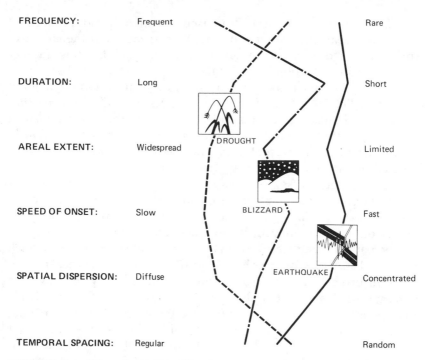

FREQUENCY:	Frequent		Rare
DURATION:	Long		Short
AREAL EXTENT:	Widespread		Limited
SPEED OF ONSET:	Slow		Fast
SPATIAL DISPERSION:	Diffuse		Concentrated
TEMPORAL SPACING:	Regular		Random

FIGURE 2.4. Hazard event profiles for characteristic drought, blizzard, and earthquake. It is possible to draw a profile for natural events with hazard potential and to make comparisons between events by characteristics independent of their human impact.

more random. A blizzard in a continental steppe region is intermediate between the earthquake and drought extremes, except in frequency and temporal spacing.

Evidence from a number of empirical investigations suggests that there is a basic difference in hazard events between the pervasive type and the intensive. Drought is the archetype of the pervasive hazard; tornado of the intensive hazard.

Pervasive–Intensive Continuum

The differential impact of pervasive and intensive hazards is illustrated by urban drought and tornadoes in the United States (Kates, 1975). Urban drought, otherwise a reduction in the natural availability of water for city needs, is a pervasive phenomenon that may affect half of the U.S. population drawing water from surface systems. The cities involved, which span the continent and include

New York, Boston, and San Francisco, together spend about $200 million per year for storage structures (reservoirs and their control works) designed to erase the normal water-availability fluctuations resulting from variations in rainfall.

A study of the effects of the drought in the northeastern United States in 1962–1966 (Russell, Arey, and Kates, 1970) showed that when there is insufficient rainfall to replenish the reservoirs in a succession of years, the cumulative deficit can lead to obviously severe shortages of water. Yet, the economic and social losses inflicted upon communities are surprisingly small — in fact, not exceeding $5 per person annually. A greater shortage would have generated greater damage per capita: the relation of loss to shortage is non-linear. If drought of the 1962–1966 magnitude has little impact on urban communities in the United States, it is because society pays a steep price in advance to protect itself from such impact. Reservoirs are designed with considerable overcapacity, partly in order to realize scale economies and partly to provide a high margin of safety. The slow onset of drought conditions permits corresponding reductions in water use by (at first) voluntary and then, if necessary, mandatory means, without causing economic or social dislocations. Thus, the chief impact of the drought hazard is seen in the steps taken to reduce its potential effects substantially.

Tornadoes, by contrast, are intensive hazards that threaten about 40 million people living in the areas of higher tornado incidence in the Midwest, the Great Plains, and the Gulf States. Tornadoes are comparatively rare events with high energy outputs and are highly localized, with a very rapid onset. Moreover, although between six and seven hundred tornadoes occur every year, the average path is quite small. There is little incentive, therefore, to invest in protective measures, since their likelihood of being needed in any one place is small and, given the force of a tornado, such measures are often not effective (Baumann and Sims, 1972). Approximately $5 million a year (compared with $200 million for urban drought) is spent to provide and improve warning systems and to construct tornado shelters or storm cellars. Warning time under the best conditions is not more than one hour ahead of the event. On the other hand, although deaths from tornado have been reduced in recent years, the number is still relatively high, averaging about 120 a year as against none from urban drought. Economic damage from tornadoes averages about $125 million a year, as opposed to only $15 million for drought.

The average annual cost (loss plus adjustment cost) per capita of urban drought for those in the area of risk is about $2; for tornadoes it is about $4. These costs are differently distributed. Drought, as a pervasive hazard, is frequent, low in energy output, slow to develop, and has stimulated a social response of high level of expenditure on prevention of effects: the ratio of sustained loss to the cost of preventive action is approximately 1:13. For tornadoes the ratio of loss to the cost of prevention is approximately 20:1.

The pervasive–intensive distinction is useful in cross-hazard comparisons. Given a knowledge of the extent to which a type of hazard or a single-hazard event is more nearly intensive, it is possible to predict the types of social response that are most likely to be adopted or to prove futile. There remains the question of whether understanding of response to a tornado makes it possible to foresee with any greater clarity what the likely response to an earthquake warning would be — if and when warnings were to become practicable.

The pervasive–intensive differences are not mutually exclusive categories but rather form a continuum along which different types of events or particular events may range. Natural hazard events falling toward the intensive end of the scale are earthquakes, tornadoes, landslides, hail, volcanoes, and avalanches. The more pervasive hazards include drought, fog, heat wave, excessive precipitation, air pollution, and snow. Air pollution is included as a natural hazard because it illustrates a situation where the hazard is largely of human origin but where natural events such as wind and temperature inversion may contribute to severe episodes.

Other hazards are less susceptible to grouping. Examples can be found at both ends of the spectrum, as in the case of floods. These might be described as complex hazards displaying mixed characteristics. Flash floods in small, mountainous watersheds are close to the high-energy, localized effects of the intensive end of the scale; whereas floods on the mainstream of a great river are more pervasive. Similarly, tropical cyclones differ widely in their impact between those affecting a narrow coastal strip subject to wave action and storm surge and those affecting the interior areas, where heavy rainfall and high wind occur but without the action of the sea. A hurricane may thus be more intensive at one point and more pervasive at another. Other complex hazards include extreme wind, glaze storm, tsunami, sandstorm, and dust storm.

An analogous intensive–pervasive distinction may be made among social hazards. Thus, the accident that in a few traumatic minutes brings disaster to a family contrasts with the grueling ordeal of long unemployment in much the same way that a tornado contrasts with drought.

Cumulative Hazardousness of a Place

Much of the public attention focused on natural hazard is channeled according to the categories of common natural hazards listed in Table 2.1, which is based upon either causal agents or common professional and disciplinary divisions. Thus, as mentioned earlier, hydrometeorologists study floods, and seismologists and earthquake engineers study earthquakes. The same professional distinctions are often carried through to the operational agencies of government; each agency or sub-unit tends to concentrate on its own set of hazard events, with the conspicuous exception of the emergency relief organizations, which routinely deal with any and all disasters regardless of their origin. There has so far been relatively little attempt to develop policies for managing natural hazards as a set of like phenomena (see Chapter 6).

The development of integrated cross-hazard policies, to be effective, requires more than the characterization and classification of hazard events described in our presentation. It requires also the correlation of such information in terms specific to a particular place or community.

London, Ontario

At two places in North America—London, Ontario, and Los Angeles, California—studies have been made of the multiplicity of hazards confronting the local community. London, a city of about 200,000 inhabitants, is located on the Thames River in south-western Ontario. The river has periodic floods that can be aggravated by ice jams, and the city is on the fringe of the heavy snowfall belt that lies to the east of Lake Huron. Winter hazards are complicated by the threat of glaze storms. Tropical cyclones from the Gulf of Mexico can reach the region in late summer and autumn. Hailstorms occur relatively often, drought very rarely, and both pose a threat-mainly to agriculture. Severe convectional storms are common in summer, and tornadoes occur infrequently.

On the basis of an analysis of historical records of occurrence (Hewitt and Burton, 1971), it is estimated that over the next 50 years in the southwestern Ontario region there may be 39 tornadoes, 25 hailstorms, 16 floods, 10 glaze or ice storms, 8 hurricanes, 5 severe snow storms, 2 severe wind storms, and 1 drought (Figure 2.5). This expectation is regional and the gap is considerable between the statistical occurrence of extreme events and the threat they pose to an individual at a particular place. For example, the

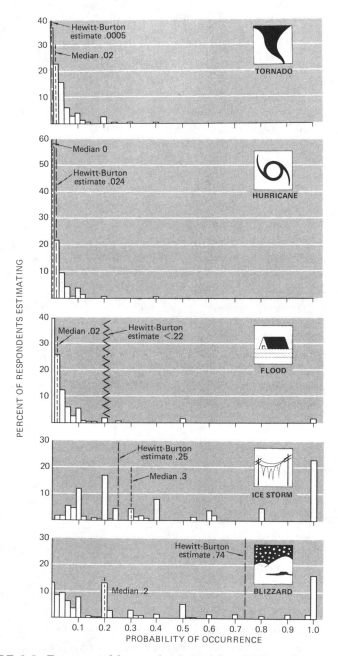

FIGURE 2.5. Experts and laypeople: Appraising the hazardousness of London, Ontario. Histograms of estimates made by residents of London, Ontario, as to the incidence of a given hazard event in 50 years expressed as a probability. The median estimate is compared with the estimates made by Hewitt and Burton on the basis of historical records.

most common category of events is that of tornadoes, which are expected to recur about every 1.3 years within the region. Yet in terms of specific urban area the recurrence probability is estimated at 400 years and, in terms of a specific house site, less than once in 2,000 years.

The cumulative threat posed before an individual household depends on the expected length of tenure. An average young household might reasonably expect to encounter at least one great snowstorm, one tropical cyclone, perhaps two floods, and three or more severe ice storms. The remote possibility also exists that the household may be hit by a tornado or a tropical cyclone stalled over the Thames basin, or by an ice storm followed by heavy snowfall — all events with potential for creating a catastrophe. In the aggregate, such events in nature clearly do not seriously threaten an individual inhabitant of London, Ontario.

Hazards of man-made origin in London are on about the same level of potential magnitude as natural hazards. At most, six disastrous natural events have occurred during the past century. This compares quite closely to the record of fire, explosion, and accident. Comparison with some 57 other North American cities of similar size suggests that the London experience is reasonably typical of North America: small, multiple probabilities of risk for the individual; community disaster from nature and man occurring almost equally each 20 years; and catastrophe recurring surely, but unexpectedly, somewhere on the continent each and every year.

Los Angeles, California

From the perspective of the whole London community, the integrated study of all hazards provides a basis for deciding what actions are appropriate either to offset the impact or to prepare for emergencies. The estimates rest, however, on the assumption that historical experience is a guide to the future. Yet, the pattern of hazard changes with time, even when the total hazardousness of a place remains constant. This phenomenon is reported in metropolitan environmental hazards in Los Angeles (Arsdol, Sabagh, and Alexander, 1964). Three hazards of largely natural origin (brush fire, landslide, and flood) and two hazards of largely man-made origin (airplane noise and air pollution) were examined. (Brush fire may of course be caused by people as well as by lightning.)

The Los Angeles study describes the spatial and demographic extent of each hazard. The most extensive air pollution (smog) concentrations are in the valley areas containing freeways, where they affect about one-half the county's 6 million people. Aircraft noise areas are related to air traffic, and holding patterns affect about one-fifth the population. Both freeway and airport areas are central and densely populated. Residential areas with high brush-fire and earth-slide potential are in dry, mountainous suburbs; slides often occur after winter rains on land denuded by autumn fires. Some 3–5 percent of the 1960 population was exposed to this hazard. Floods are scattered throughout areas affecting about one-sixth of the population.

Significant changes have taken place over time in the mix of hazards threatening Los Angeles, while at the same time the total proportion of the region exposed to hazards has declined. As suburbanization has taken place, the relative proportion of the exposed area to the central city noise and smog has declined, to be replaced by new amenity-related hazards of slide, fire, and flood (until 1950) — a pattern characteristic of urban expansion in North America.

RESPONSE TO HAZARDS

Response to hazards is related both to perception of the phenomena themselves and to awareness of opportunities to make adjustments. Rarely are individuals unaware of the existence of possible hazards, yet their perception and definition of the threat may differ markedly from the estimates of professionals and experts.

In the study of the hazardousness of London, Ontario, de-scribed above, for the first time an effort was made to compare in a systematic way "lay" appraisals with more "objective" assessments across a range of hazards. The lay perceptions were based on interviews in which London residents were asked what natural hazards they believed threaten the city (Moon, 1971); the objective estimates, drawn from Hewitt and Burton, were based on historical records.

Respondents in London showed a moderately high degree of awareness of the five major hazards; two-thirds of the responses designated them. As for expectations of the frequency of major hazard (respondents had been asked how often a given event could be expected to occur during the next 50 years), responses are plotted in Figure 2.5, together with the more nearly "objective" estimates.

Events were believed by the respondents as more likely to occur 5, 10, 15, or 20 times in 50 years rather than at odd intervals such as 4, 7, 9, 11, 14, or 16. A preference for multiples of 5 is to be expected; too, the distribution of responses for tornado, hurricane, and flood coincides with observed frequencies. That 140 people are persuaded that no flood will occur within the next 50 years very likely reflects the high degree of confidence placed in the Fanshawe Dam, which does reduce flood peaks. Most significant about tornadoes, hurricanes, and floods, however, is that the subjective estimates of frequency fall close to each other and are relatively close to the observed frequencies.

The expectations of occurrence of ice storms and blizzards are quite clearly in a different category. Both have a bi-modal distribution, with marked peaks toward each extreme. Ice storms are expected to occur between 1 and 5 times in 50 years by 86 people and between 46 and 50 times by a somewhat smaller number. Similarly, blizzards are expected to occur 5 times or less in 50 years by 148 people, and about one-third that number expect blizzards to occur 46 times or more in 50 years. This suggests a discrepancy in the way different people define ice storm and blizzard (most dramatically in the case of blizzards, people may view them either as common events occurring almost every year or as rare events occurring only a few times in 50 years). In neither case do the peaks of subjective frequency coincide with objective estimates based on the observed record. On the contrary, the objective frequency tends to fall into an area of very low subjective expectation (Burton and Moon, 1971).

The significance of these observations is that they strongly inhibit the propensity of individuals or households to adopt adjustments to specific hazards for which the perceived frequency is low. Differing concepts of what a hazard signifies may also affect the receptivity of populations to warnings or other advice about adjustments.

The Range of Response

A theoretical range of response encompasses all the ways in which a society may act to reduce the effects or increase the benefits of a hazard. These may grade from immediate actions in the face of danger (such as warning systems, flood fighting with sand bags, or emergency evacuation) to long-term actions (such as planting crops less susceptible to drought or the construction of buildings designed to resist earthquake). Other responses include the long-run adaptations of a culture to the extremes of its environment, as in building villages on levees. Many actions are taken that have the incidental effect of reducing vulnerability to hazard, such as the use of construction materials of lesser susceptibility to earthquake damage solely because in some places they are cheaper. Over the very long run, mutations may occur in human biology, as in capacity to control body temperature. The timing of these responses is shown in Figure 2.6 relative to the occurrence of a hazard event.

Building a dam to store additional water for irrigation during a drought period would be classed an adjustment. A system of cut-and-slash farming in the Laotian highlands, with all its requirements for appropriate social organization, in timing the cutting, burning, cultivation, and revegetation of forest lands, would be counted an adaptation. Designing a house to resist a storm surge would be an adjustment; locating and organizing a community over a long period of time so that its houses are beyond the reach of storm surge would be an adaptation. An individual or group may, however, choose to apply as an adjustment a practice that long has been an adaptation elsewhere, as when a home owner builds a house flood-proofed with a design imported from a distant place.

Biological Adaptation

In evolutionary theory many physiological and anatomical characteristics of the human species, as well as of other species, are thought to result from a "natural selection" process in which those traits most advantageous for survival and success in a given environment were selected and maintained. This was presumably the case in the emergence of Homo sapiens, and the process of biological adaptation doubtless continues. Although the role of extreme events and environmental changes in the evolutionary process is far from clear, it does not appear to have been a leading one so far as geophysical events are concerned (see Burton and Hewitt, 1974). The process of biological adaptation is generally slow; it cannot play a significant role in the short-run response to natural hazards except as population traits may be revised through exposure to disease, through measures to suppress disease or maintain affected individuals, and through changes in nutrition.

The capacities and limitations of the human body are nonetheless extremely important in setting constraints on other kinds of response. The susceptibility of the human organism to environmental hazards is an increasingly urgent question in public health and preventive medicine, and much more research effort (not considered here) is now being turned in this direction.

There are, however, very few biological adaptations that can be related specifically to short-run events. More often they can be

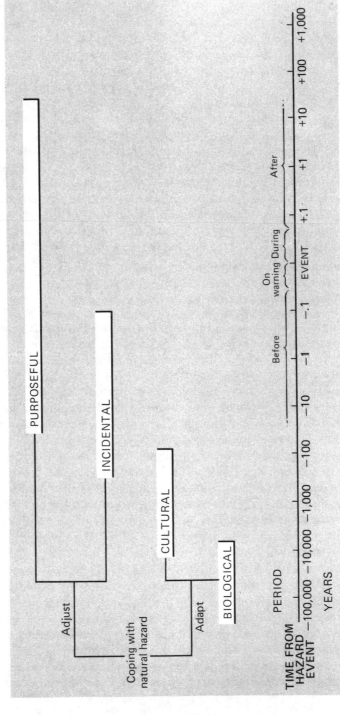

FIGURE 2.6. Time scale for coping with natural disasters. The four major types of response or ways of coping with hazard involve very different time scales in order to affect a hypothetical event.

linked to environmental conditions or states rather than short-run environmental changes. Thus, the mechanism for redistributing heat among parts of the body, especially developed in Australian aborigines, is a response of either genetic or physiological character to the conditions of the environment and the level or type of available technology. Similarly, the existence among many Africans of the sickle-cell trait, which gives a degree of immunity to malaria, is a biological adaptation that developed over long periods of exposure to the disease. Biological adaptations may also involve numerous mechanisms for temporary physiological responses in the face of hazards; one such is the capacity to increase adrenalin levels rapidly. These mechanisms are excluded from the present framework, as is the possibility—not yet realized but potentially attainable—of controlled evolution by man-manipulated biological change.

Cultural Adaptation

Cultural change can be a much more rapid process than biological adaptation and has in fact accelerated greatly in the last two centuries. With respect to natural hazards, the effects of cultural change have not necessarily been in the direction of greater adaptation.

In traditional societies, such as the Innuit of the Mackenzie River delta or the Fulani herdsmen of the Sahelian zone on the southern fringes of the Sahara Desert, the process of cultural adaptation may result in relatively stable relations between people and their environment. When this slowly shifting balance is upset by external forces that generate rapid cultural change, new and unsuspected natural hazards can appear. For example, the massive transformations likely to follow the discovery and exploitation of oil in the Mackenzie delta may have a profoundly destructive effect on delicate Arctic flora and fauna, thus exposing the native populations to new threats arising from instability in the traditional environmental supports of fish and game. Similarly, in the Sahelian zone the introduction of improved domestic water-supply facilities encouraged growth in the number of livestock well beyond the capacity of the semi-arid grasslands to support in times of natural moisture deficiency. As herds enlarged and pastures were degraded and population expanded, the society became more vulnerable to dislocation when drought struck, and the marketing and other economic measures taken by governments caused severe distress to cultivators as well as herdsmen.

The adaptation may also involve ingenious and beneficial use of the available resources. Cultural adaptations of interest here embrace the many ways in which human activities are structured that have the effect of reducing or enlarging damage from extreme natural events. As they are shifted they can lead in directions that diminish the exposure of human populations to environmental hazards, or they can engender an expansion of risky activities.

Adjustments

When Mackenzie Innuit or Fulani herdsmen are subjected to environmental conditions that stretch their society beyond present adaptations, they can make a wide variety of adjustments in their ordinary activities. Once adopted, a new adjustment may be short-lived, or over time it may become an integral part of the adaptive fabric of the culture. Responses to hazard that are capable of short-run adoption generally offer the earliest and most promising results in reducing losses and saving lives. In a set of comparative responses to natural hazards in fourteen countries reported elsewhere (White, 1974), many residents of hazard zones were asked what could be done about natural hazards. The replies to such questions generally referred to a limited range of actions that could be taken by individuals, small household groups, communities, and national governments. Compared with cultural adaptations, these actions usually are put into effect relatively quickly and may be only temporary. They vary greatly in number according to cultural setting and type of hazard. Respondents in northern Nigeria could only suggest a few adjustments to drought (see Table 2.2) — much fewer than in Tanzania. In contrast, 264 separate adjustments to floods were suggested in Sri Lanka.

Adjustments may be separated into those that are purposefully adopted and other activities and characteristics of individual behavior that sometimes are not primarily hazard-related but have the effect of reducing potential losses. Such adjustments are described as incidental. Thus, improvement in the quality and strength of building materials has reduced vulnerability to high-impact hazards in developed countries, although this has not been the purpose of the change. For example, brick houses are much less susceptible to tornado damage than those of wood. Advances in communication systems, including radar, satellite observation, and radio and telephone links, have provided previously unavailable opportunities for hazard response that are incidental to their main purposes. Increased capacity to store surplus grains and to move grains from

TABLE 2.2. Adjustments to Drought Suggested by Peasant Farmers in Nigeria and Tanzania

Northern Nigeria		Tanzania
Nothing permanent	*Change location*	Nothing permanent
Nothing	*Change use*	Drought-resistant crops
		Irrigate
Store food for next year	*Prevent effects*	More through weeding
Seek work elsewhere		Cultivate larger areas
temporarily		Work elsewhere
Seek income by selling		Tie ridging
firewood, crafts, or grass		Planting on wet places
Expand fishing activity		Sending cattle to other
Plant late cassava		areas
Plant additional crop		Sell cattle to buy food
		Staggered planting
Consult medicine men	*Modify events*	Employ rainmakers
Pray for end of drought		Pray
Turn to relatives	*Share*	Send children to kinsmen
Possible government relief		Store crops
		Government relief
		Move to relative's farm
		Use savings
Suffer and starve	*Bear*	Do nothing
Pray for support		

Sources: Hankins (1974); Heijnen and Kates (1974); Dupree and Roder (1974).

place to place over considerable distances likewise reduces vulnerability to low-impact or pervasive hazards such as drought. A house builder along the Rocky Mountain piedmont may decide to use building designs that will prevent wind damage. The same engineering features will also reduce earthquake damage—an eventuality for which few builders in that area provide and of which the owner may be unaware.

The classification of responses suggested in Figure 2.6 does not provide mutually exclusive categories. In particular, there is considerable overlap between purposeful adjustments and cultural adaptations. As already noted, a response initially adopted as an adjustment, either purposeful or incidental, may gradually become transformed into a cultural adaptation. The break point at which a pattern of land use ceases to be thought of as a response to hazard and becomes the way things are habitually done is not itself important. Rather, the importance of cultural adaptations and incidental adjustment lies in the way they act upon a society's absorptive capacity for natural hazard.

Absorptive Capacity

When hazards are seen as a function of both natural events and human-use systems, then just about any change in either, however small, can be shown in theory to have some potential impact on damages. It is not practicable, however, to assess the whole human-use system in terms of its effects on damage potential from natural hazards. The task, extending into every aspect of social and economic life, would be too large and cumbersome. Attention focuses primarily on purposeful adjustments. Nevertheless, incidental adjustments and cultural adaptations are of basic importance inasmuch as they create the level or capacity of individuals, managerial units, and social systems to absorb the effects of extreme environmental fluctuations.

For example, the Kamba farmers at Kilungu in Kenya practice a method of cultivation that appears to the external observer rather jumbled and disorderly. Crops are interplanted with no apparent semblance of order. The complex schedule of planting, harvesting, and second planting gives the fields an appearance of chaos. The practices involved, however, include many ingenious and adaptive elements that help to ensure a moisture supply for the crops. A farmer may plant maize, beans, cow peas, sorghum, ground nuts, bullrush, millet, red millet, cassava, pumpkins, callabashes, and pidgeon peas all mixed together in one field. The people of Kilungu do not describe this practice as an adjustment, but to plant physiologists and agroclimatologists it has much merit. According to one investigator,

> The crops come up altogether in a riotous profusion of vines, leaves and stems. Weeds don't have a chance. The phosphorus flush in the soil, which comes with the first rains, is taken advantage of by these crops. Crops such as maize, which might be rather indolent about putting down roots and strong tillers in a field well supplied with moisture, have to compete with other plants and thus they put down roots to a depth of several feet. These deeper roots come into play later on in the season. Although the rate of moisture use of all these crops planted together is high, it occurs at a time when there is commonly plenty of moisture available. Further, it provides a good continuous canopy within which photosynthesis can proceed. Since tropical areas have a relatively uniform radiation income all year round, the continuous leaf canopy, one with a moderate to high leaf index, is the most efficient user of radiant energy for photosynthesis.
>
> After about seven or eight weeks some of the crops are harvested. These may be harvested as green vegetables. As the season progresses

the millet is harvested, beans harvested for seed are taken and the vines pulled up. The number of plants per square meter begins to decline and only crops requiring a longer time to mature are left. The bare weedless soil between the plants is dry, which forms a barrier to the movement of moisture to the surface from below. Evapotranspiration thus is reduced. The sparse plant population remaining is able to tap moisture from a larger volume of subsoil without competition from adjacent plants. This moisture, combined with the lesser amounts of rain which come at the end of the grass rains, is sufficient to bring the maize and the longer growing millets to harvest. The plants also provide some shade protection for crops set out to get a start on the main rains. The thick mat of plant cover in the first weeks of the grass rains and main rains also serves to hold the soil. There is also the adaptive fact that if the rains do give out, some crops will have been harvested and the agricultural effort will not have been a total loss. Another fact is that interplanting reduces the amount of work considerably by eliminating the need to weed, which often is a most serious impediment to agriculture and the management of larger acreages. (Porter, 1970)

Clearly the agricultural system has considerable built-in capacity to absorb the effects of drought. The practices might be described as cultural adaptations in the long term or incidental adjustments in the short term. In either case they are largely unconscious on the part of the farmer. In a sense he farms better than he knows. The practices have been developed over a long period of trial and error. Replacement of such a system by a more commercially oriented agriculture, with crop segregation instead of interplantings to facilitate the introduction of mechanization, might make for greater efficiency and perhaps higher income at substantially greater investment in the short run. It would very likely increase vulnerability to drought. Modernization of agricultural practices at Kilungu without careful safeguards and extra expense would probably reduce absorptive capacity to drought and create a potentially disastrous situation.

The concept of absorptive capacity applies as well to industrialized nations and may be illustrated by reference to municipal water systems in the United States. Stream flow is highly variable, and water supply systems dependent on it generally require some storage capacity to provide for the daily, seasonal, and year-to-year fluctuations in both supply and use. The basic adjustment is reservoir storage. A measure of the adequacy of storage may be devised by comparing the level of water use to safe yield, a measure of the reliability of supply (conventionally defined as the amount of water that river flow with available reservoir storage can provide 95 percent of the time). This measure of reliability also indicates part of the absorptive capacity of the system. Where water use is much below safe yield, meteorological drought can usually be accommodated, given the high margin of safety, without causing shortages. Conversely, if water use exceeds the safe yield by even half (150 percent), such a system is vulnerable to even slight variation below normal flow.

In comparing data on the percentage of safe yield used by systems serving populations in different climatic regions of the United States, substantial differences in percentages are found between regions. In the Southwest the median system's water use in 1962 was 41 percent of safe yield, in the Midwest, 60 percent, and in the Northeast, 73 percent. These are related to climate, mainly the interregional differences in the variability of precipitation and stream flow. Such variability is greater in the Southwest than in the Midwest, and greater in the Midwest than in the Northeast. The managers of municipal water systems in the Southwest have incorporated a greater level of absorptive capacity into their systems.

It is easier to specify in outline the concept of absorptive capacity than it is to propose common measures for its size. The phenomenon of the safety margin is widespread, and judgment about its necessary size will depend, among other things, upon the view of hazard that is held. Describing subsistence cultivation in Africa, Allen (1965) uses the concept of normal surplus. In the face of larger year-to-year variability in crop production (related in considerable part to precipitation), subsistence cultivators produce what they need in a below-average year, thus providing a surplus in above-average years, and some surplus in normal years. Overcapacity in production is the norm, even though the surplus is not easy to dispose of in good years and requires considerable effort to create.

The farmer may be obliged to strive for a surplus in order to assure his own economic security, or to protect himself against exploitive credit demands in case of crop failure, or to meet national demands for a cash export. The situation will vary with the prevailing political system: the threat of crop loss for a tenant laborer in a landlord-dominated economy is different from that for a member of a commune. Likewise, the political context in which a water-system designer plans for the (rare) dry year varies with the government policy. What is an acceptable risk in one country carries heavy professional or government sanctions in another if shortages develop.

Operators are sensitive to the economic cost in maintaining this excessive productive capacity. A measure of how much the surplus might be was obtained by asking farmers in three moisture areas of the Usambara Mountains of Tanzania what their remaining surplus of grain amounted to just before the new harvest in a good year. For farmers in the high-moisture zone the median estimate was about 120 kilograms, or about one month's supply. In the moderate and low-moisture zone, surpluses almost 3 times greater are common. In drought-hazard situations as diverse as those faced by North American urbanites and Tanzanian smallhold farmers, the evidence is that the demand for considerable absorptive capacity may increase with the variability of environmental events.

The question of how much absorptive capacity is needed is now being recognized for the first time on a global scale, especially in relation to the supply of food grains. Recent years have seen massive shipments of grain from North America to the then Soviet Union, China, and India in periods of crop failure due to floods or droughts. Given the risk of simultaneous crop failure in several of the major grain-producing regions of the world (and the remote but possible event of failure in all of them), how much should be spent in the development of an internationally available grain surplus? How much absorptive capacity exists and should exist on a global scale for food grains?

A Choice Tree of Adjustment

One way to visualize the large range of choices available to individuals or agencies is as a tree of alternatives involving a scale of increasingly active and complex adjustments. Such a tree of possible choices of adjustments suitable for individuals and small groups is shown in Figure 2.7. The basic decision concerns livelihood and location — that is, choosing a certain activity at a particular site. For many individuals and groups, particularly in less mobile societies, that decision may seem fixed — land is incidental to maintaining their livelihood — but such is rarely the case. Any locational decision is initially made as purposeful choice, and most societies do provide rare opportunities to renew or shift commitments. Inheritance is one such opportunity; disaster frequently is another. Within a given location a range of livelihood choices may be open, and small or large shifts in the pattern of use can be made.

Once located and committed to a particular resource use, people use a variety of psychological, personal, and social devices to (1) discount losses by disregarding them or including them with

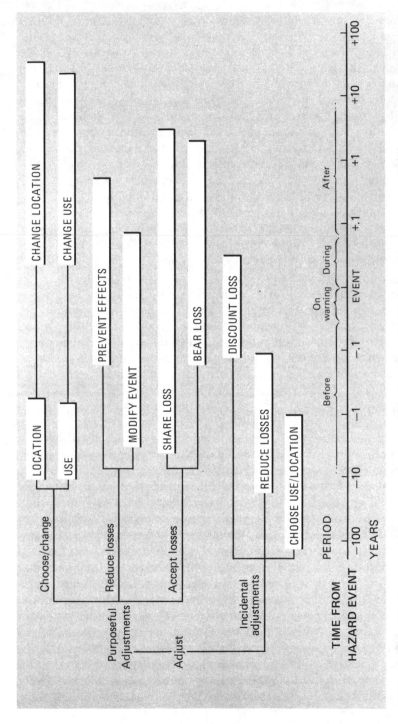

FIGURE 2.7. A choice tree of adjustment. Adjustment begins with an initial choice of a resource use, livelihood system, and location. For that choice various incidental and purposeful adjustments are available, at somewhat different time scales for initiation. The most radical choice is to change the original use or location.

other costs of location, or to (2) accept losses or to distribute and
share them with other people. Whether people seek to reduce losses
only after the occurrence of a hazard event or during the event (with
a warning of possible occurrence), or much before the event, reflects
increasing commitment of effort, forethought, and choice. They
reduce losses by reducing the damage potential or by modifying the
events themselves. As more and more substantial actions are seen to
be required, the inhabitants of hazard zones may again be confronted with the choices of changing use in place or of migrating to
another place. By no means are all adjustments to hazard purposeful, nor are all the theoretically available purposeful adjustments
necessarily employed in a given place in relation to a given hazard.

Purposeful Adjustments

The lexicon of purposeful adjustments for hazard in a human use
system at any given time can be compiled by systematic survey of
those involved. These include individuals and household groups,
communities, the managements of commercial or industrial enterprises, and regional and national governments. From the large
theoretical range of purposeful adjustments, the portion actually
used constitutes a distinctive mix for each location. One interesting
aspect of collective endeavor in the People's Republic of China is
that it encourages systematic investigation of folk wisdom having
possible application to modernized agriculture, and thus consciously
enlarges the range of adjustment options.

Each purposeful adjustment can be ranged along a time scale
centered upon the occurrence of the extreme event. As diagrammed
in Figure 2.7, some adjustments, such as changes in land use or
insurance against damage, can be effective only if begun well in
advance of the event. Others can be initiated only as the damage
occurs or in its wake, and may serve to reduce impacts either during
the period of reconstruction or at a later time.

Accept Losses

People who accept losses without attempting to alter the causes may
either bear the full burden of damage or share it in some fashion
with others. Those who *bear the impact* independently may not pass
the losses on to others in overt ways: their income may be
diminished or their lives disrupted, and they accept the consequences. When a manufacturer locates his new plant in what he
knows to be a hazardous zone, he may be deciding that this is the

alternative most likely to be remunerative to him, and if there are losses he will try to bear them without outside help.

Even when individuals or communities choose to bear the loss, there may be social devices to *share the burden* that sooner or later will fall on them. Insurance schemes are a formal way of spreading hazard, and always carry the additional cost of administration, which may fall directly on those who pay premiums or, when it is organized by a government, indirectly on all taxpayers. Coverage by insurance may be wholly voluntary or wholly mandatory, and large businesses often self-insure by setting up reserves and diversifying. Relief and rehabilitation assistance may be offered disaster victims by both public and private agencies, and in many societies a peasant looks to relatives and friends rather than to any government bureau or voluntary social agency for help in time of need. But beyond these arrangements, societies may make intricate provisions to offset disaster losses: festivals, family ceremonies, labor sharing, and a variety of other devices have the effect of aiding the sufferer from natural events.

Reduce Losses

The dramatic but rarely effective measure is to *prevent the event* from occurring in the first place. Beyond the traditional exercise of magical powers to prevent a tropical cyclone or a flood or a locust plague, modern technology has experimented with a few measures such as cloud seeding, which has been used to dissipate hail storms or to increase precipitation. Thus far, most of these are of limited application, and most, among them earthquake prevention or hurricane dissipation, are still in early stages of development.

The more common mode of reducing loss is to alter the vulnerability of society to the hazardous event by curbing it or by designing human activities to *prevent its injurious effects*. The principal types of activities are

- warning systems, comprising forecasts, dissemination of warnings, emergency evacuation, and preparation to take effective action upon receipt of the warning, such as flood-warning services.
- control works, including coastal barriers, levees, water storage projects, and earth stabilization.
- building design and construction to resist wind, water, and earth movement, such as elevation of new buildings in a hurricane zone.
- cropping practices to render production more reliable.

Preventive adjustments require considerable lead time for planning and execution before the event occurs, and some adjustments must be initiated many years in advance. Scientific studies to lay the groundwork for a hurricane-forecasting network may consume decades, and even the construction of a small storage dam to supplement irrigation water supply for a few hundred hectares of semi-arid land ordinarily demands 10 years from the initiation of design until the project is fully operative. Although a few measures, such as emergency evacuation or salvage planting of cassava after other crops have failed, can be undertaken on short notice without undue costs, a period of planning and substantial expenditure usually is involved.

Choose Change

Purposeful acceptance or reduction in loss usually assumes that people will remain in the same place, gaining a livelihood in substantially the same way, oftentimes at heavy social cost. The alternative possibility is to change the basic pattern of production or to migrate to another location.

Examples of *change in use* are shifts from dry farming to irrigated farming, substitutions of manufacturing storage yard for machine shops in a floodplain, or replacement of residences by park facilities on a strip of hurricane coast. The land remains productive but with a different pattern of use. When this involves the transfer of individual households or business enterprises, it replaces them with other economic activities.

In contrast, a *change in location* brings wholesale migration of communities or sub-groups. Thus, the Brazilian farmers who move permanently to the city, or Sahelian herdsmen who take up sedentary agriculture when drought destroys their livestock, are permanently changing their location. In the new location the migrant faces the same theoretical range of purposeful adjustments to the hazards prevailing there as in the former location. The person who doesn't move always has that extra option of migration. It is uncommon for people to rely wholly on one adjustment, and it is rare that all the people in a community adopt the same mix of purposeful activity.

CHOICES AND DECISIONS

When confronted with a hazard event, a farmer, a plant manager, or a government agency may simply repeat the protectionary and precautionary actions taken on previous occasions. They may begin a more systematic canvass of possible alternative adjustments, including any new options that may be found or created. In any event, a choice is required.

Methods of Choice

How do people choose the degree of risk they will bear and the adjustment they will make? Much of the process of judgment and choice still is the object of speculation. There is a good deal of economic and psychological theory to suggest the process of choice. Some of these theories are substantiated, more are not; and they are not mutually consistent. Most of the models were developed for normative rather than descriptive purposes.

In technical parlance, the person who chooses on the basis of assessment of all the expected outcomes is using *expected utility methods*. The person who chooses not upon the basis of probable outcomes but on his subjective view of those probabilities is using *subjective expected utility methods*. The person who arrives at judgment of the best course of action by subjective assessment of utility, with something less than the goal of choosing the maximum incremental returns, is using *bounded rational methods*.

There are other views of the way people arrive at their choices. If it were known precisely what processes people follow in choosing among the options open to them, it would be easier to isolate the factors influencing their final choice. In the absence of such confirmed theory, it will help to illustrate three of the more common views by the relatively simple case of a fisherman who has heard a storm warning.

Fisherman's Choice

The choices are usually sharply limited for a coastal fisherman who learns from studying the skies or from listening to a government radio broadcast that a tropical cyclone soon may either sweep across his stretch of Caribbean beach or cross the beach elsewhere. To take the simplest case, he may either flee inland with his equipment in the few hours open to him, or remain in his shelter and hope to sit it out if it comes. This choice is described in Figure 2.8 as that between evacuating (A_1) and staying put (A_2). The capricious natural event foreseen also may be stated most simply either as a state in which the hurricane strikes (E_1) or one in which it takes another path (E_2).

FISHERMAN'S CHOICE

A. EXPECTED UTILITY: COMPLETE UNCERTAINTY

STATES OF NATURE

ALTERNATIVE ACTIONS	E_1 Hurricane	E_2 No. hurricane
A_1 Evacuate	Equipment intact Pay for evacuation (+1)	Equipment intact Pay for evacuation (+1)
A_2 Remain	Lose equipment (0)	Equipment intact (+2)

B. EXPECTED UTILITY: KNOWN PROBABILITY

STATES OF NATURE

ALTERNATIVE ACTIONS	E_1 Hurricane	E_2 No. hurricane	Expected utility
A_1 Evacuate	Equipment intact Pay for evacuation .4(+1) = .4	Equipment intact Pay for evacuation .6(+1) = .6	1.0
A_2 Remain	Lose equipment .4(0) = 0	Equipment intact .6(+2) = 1.2	1.2

C. EXPECTED UTILITY: SUBJECTIVE PROBABILITIES

STATES OF NATURE

ALTERNATIVE ACTIONS	E_1 Hurricane	E_2 No. hurricane	Expected utility
A_1 Evacuate	Equipment intact Pay for evacuation .9(+2) = 1.8	Equipment intact Pay for evacuation .1(+2) = .2	2.0
A_2 Remain	Lose equipment .9(0) = 0	Equipment intact .1(+4) = .4	.4

D. EXPECTED UTILITY: MINIMIZE REGRET

STATES OF NATURE

ALTERNATIVE ACTIONS	E_1 Hurricane	E_2 No. hurricane
A_1 Evacuate	Equipment intact Pay for evacuation −1	Equipment intact Pay for evacuation −1
A_2 Remain	Lose equipment −2	Equipment intact 0

FIGURE 2.8. Payoff matrices for equipment evacuation on tropical cyclone warning. In trying to decide whether to evacuate the boat with his equipment or to "sit it out" in the face of a tropical cyclone warning, a "rational" fisherman might analyze his choices in many ways depending on his knowledge, beliefs, and values.

Expected Utility

If we assume that the property he could evacuate has a value of two units, and that the cost of evacuation would be one unit, the four possible outcomes in terms of their expected utility for the fisherman will be as shown in Figure 2.8. If he does not evacuate and the storm strikes elsewhere (A_2, E_2), he loses nothing. If it crosses his section (A_2, E_1), he loses all. If he evacuates $(A_1, E_1$ or $A_1, E_2)$, he loses the cost of evacuation regardless of the state of nature. Given a state of complete uncertainty as to whether the hurricane will strike, it is necessary for him to judge which alternative promises the larger return (see Slovic et al., 1974, for a summary of expected utility theory and its formal groundwork).

Let us assume that the probability of the hurricane strike can be established accurately. The probability is 4 out of 10 that the hurricane will strike. The expected utility of the choice now can be calculated as shown in Figure 2.8. In these circumstances the alternative of not evacuating will maximize the expected utility. This conclusion assumes, however, that our values and probabilities adequately represent those of our hurried fisherman.

Subjective Expected Utility

The fisherman may attach quite different value to his equipment or to the evacuation operation than would an outside appraiser, and he may also appraise the chances of a strike on another basis than would a meteorologist backed by reams of records. His subjective judgment determines the pay-off he expects from the two courses of action.

For example, he may regard his equipment as worth double the expected value, and he may count the cost of evacuation as double that expected. Upon the basis of experience with a hurricane the preceding season, he may regard the chances of its crossing his beach as 9 out of 10 rather than 4 out of 10. These reflect his considered view, and he acts upon them. The pay-off then becomes that shown in Figure 2.8. The greater probability of the extreme event (0.9) makes it seem to him far more productive to evacuate. Here, too, the question arises as to whether a subjective calculation of expected utility correctly reflects the choice process of our fisherman.

Bounded Rationality

Let us assume that the fisherman evaluates the utilities and appraises the hurricane threat in a manner that corresponds completely with the estimates given in the expected utility matrix of Figure 2.8. He knows that the "best bet" would be for him to stay in place, all other factors being equal. The other factors are not necessarily equal, however. His society may offer inducements either to flee or to remain in place.

He may not be interested in maximizing the pay-off: rather, he may be willing to follow any course of action that assures him that his costs will not exceed one unit. In that case the difference between 1 and 1.2 may be unimportant for him. Or, he may be preoccupied with keeping out of debt to the local moneylender. In that instance he may not consider pay-off primarily. Instead, he may look at possible regrets as shown in Figure 2.8. In thinking of probability the fisherman may be as much concerned with the extreme event as with all other more probable events. His primary aim may be to keep out of debt. He would therefore evacuate in the face of a hurricane up to the point that evacuation would put him deeper into debt than would remaining. A loss of two units may seem to him catastrophic and to be avoided at any lesser cost. Common choices are often made by a process that Simon called "satisficing" (Simon, 1956).

As will be shown, it is rare indeed that individuals have access to full information in appraising either natural events or alternative courses of action. Even if they were to have such information, they would have trouble processing it, and in many instances they would have goals quite different than maximizing the expected utility. The bounds on rational choice in dealing with natural hazards, as with all human decisions, are numerous. The process is stressed here because there is a tendency to expect that people left to themselves with enough information will select the optimal adjustments. Any doubt about this is dispelled by examining the kinds of experience with natural hazards provided by both developing and high-income nations.

CHAPTER THREE

The Range of Experience

Although four national comparisons do not mirror the whole variegated tapestry of response on a global scale to extreme natural events, they do throw light upon major facets of the experience. They also provide background against which to examine what is known about the processes by which individuals, communities, and nations arrive at their own particular stance toward hazards.

In certain respects the experience is similar, while in others it displays great disparities. Drought weather in Australia is much like drought weather in Tanzania, and yet the mix of adjustments practiced by farmers and their governments differs significantly. Although flood characteristics in the United States are more complex than those in Sri Lanka, the fundamental problem of responding to threats of overbank flow are similar, and the types of response on a national scale are alike in some respects and quite unlike in others. The contrast between tropical cyclone experience in Bangladesh and that in the United States is even greater. In the struggle with air pollution, the measures taken in the United Kingdom are of a greater order or magnitude than those adopted in Mexico, but they both illustrate the interaction of technology and local atmospheric conditions in producing serious hazards.

Comprehensive analysis of the hazard problem for a national territory is a recent scientific undertaking; the first such effort, one dealing only with floods, was organized in the United States less than 20 years ago. More recently, reviews were undertaken in Canada, Japan, and the former U.S.S.R. In these studies the pattern of declining deaths and rising damages, and the substantial investment in technical means of coping, were clearly discernible. The prevalence of this pattern in nonindustrialized society was not established, however; most of the information was anecdotal. A few field studies are of sufficient breadth to constitute appraisal of the national experience for a single hazard, and they serve as source materials for this chapter.

The comprehensiveness of these studies varies considerably, as being limited by the resources available to the collaborating researchers. Even in industrialized countries data routinely collected for hazard situations are sparse, and new data are hard to obtain. Field work in Sri Lanka was hampered severely by civil strife occurring in the countryside. The first field study in Bangladesh took place just before the November 1970 cyclone; the second awaited the birth of the new nation.

The ranges of areas and population covered in the national studies are shown graphically in Figure 3.1, with 20-fold differences apparent in areas and populations. They represent only a small sector of the entire range of developing and industrialized nations. Patently, agricultural drought, flood, or hurricane covers only a few

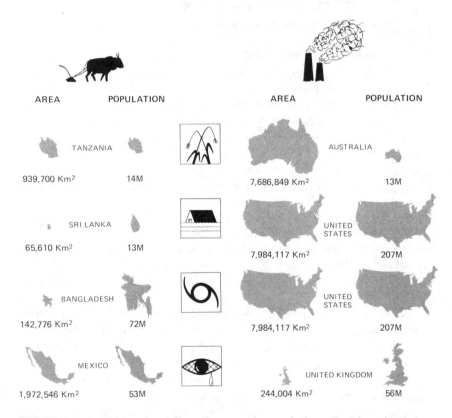

FIGURE 3.1. National studies: Area and population. For four hazards—drought, flood, tropical cyclones, and air pollution—national studies of a developing nation and an industrialized nation were undertaken. Area and population differ considerably among nations: as is shown by the proportional country outlines.

sections of the continuum of characteristics of hazard-engendering natural events. In later chapters several other hazards are discussed—such as earthquake in selected areas—and compared with what is known of other areas, some of which (such as the People's Republic of China) are known only through a brief visit or meager publications. The comparison with air pollution is at best only suggestive of what differences and similarities can be found between events primarily of natural origin and events of human origin. But these comparisons expose a bit more of the complex warp of social and physical systems upon which is woven the adaptations and adjustments people make to hazards.

AGRICULTURAL DROUGHT

Tanzania

In area larger than Britain and France together, Tanzania is home for 13 million people, 90 percent of whom work and live on the land. For them, maize, cassava, and bananas serve as major staples; cotton, coffee, and sisal are major export crops. Much of this land is semi-arid, and supports almost as many cattle as people. Since Tanzania extends 10 degrees south of the equator, with elevations ranging from the snowfields of Kilimanjaro to the shores of the Indian Ocean, its rainfall pattern is highly seasonal and complex. There are areas with distinct wet and dry seasons, and areas with multiple seasons, with both "short" and "long" rains.

Drought occurs somewhere in the country 2 out of 3 years. One year in 5 it affects upward of half the population; 1 year in 10 major famine potential is experienced. In such years as many as a million households suffer setback to their modest hopes for improved life and livelihood and, more rarely, the death and destruction of famine (see Figure 3.2).

The social cost of agricultural drought includes both the actual loss of plant and animal production, related health, nutrition, movement, and income effects, and the cost of efforts expended to prevent, reduce, or replace such losses. A crude appraisal suggests that a year of severe drought (1964–1965) would cause a loss of crop and animal production of about 8 percent valued at 3 percent of the $1.1 billion gross national product (GNP) at current prices. The order of magnitude of losses calculated by this method is given credence by a drop of 10 in the percentage assigned to value of agricultural production as recorded in the National Income accounts for 1965. Extrapolating to the 25 years between 1945–1946 and

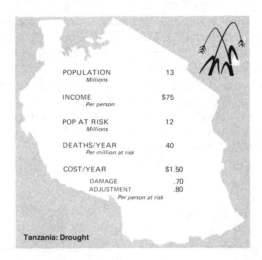

FIGURE 3.2. Tanzania: Drought.

1969–1970, average annual loss of production was about 2 percent valued at 0.75 percent of current GNP.

The impact of drought on the life, health, and well-being of the population is even more difficult to appraise but too important to ignore. A single-season drought cannot by itself account for significant nutritional impairment, though it can accentuate the vulnerability and long-term mental impairment of the very young, who already suffer death rates of 30–40 percent. Beyond the physiological stress of drought and resulting food shortage is the psychic stress ranging from the sense of deprivation that accompanies marked reduction in accustomed diet, through diminution of herds and loss of savings, to the universal grief in the face of death and catastrophe.

Tanzanians encountering frequent drought know that the rains will come again, and this knowledge is reinforced for many by a deep religious faith that sees a benevolent God as protector. The kind of desperate but futile action in the face of calamity sometimes found elsewhere is uncommon in Tanzania, for there are many constructive post emergency adjustments, useful in all but the worst of famines. When these fail, serious anxiety occurs. Moreover, the relative equality of wealth (or poverty), as well as the strong tradition of clan and extended-family sharing, prevent the painful sense of collective suffering amid individual plenty that characterizes disaster in more highly stratified societies.

If the burden of loss from frequent shortfalls of moisture

appears relatively low, it is because of a continuous effort of adjustment. In Tanzania the burden for such adjustment falls in the main on the efforts of the farm family, although over time a trend is discernible for responsibility to shift to more collective levels of social organization, resting finally on the central government.

Prayer, planting in seasonal floodplains, planting drought-resistant cassava or sweet potatoes, and buying food are the most common adjustments for the farm households. The most common costs involve labor, but may involve also payments in kind or in money to rainmakers, the risk of losing seed when planting dry, or the social-psychological stress of moving from one's family, begging help, or selling off cattle. From an examination of farm-work budgets, 10 days per year for a family of five or six is a reasonable approximation of labor invested in reducing possible drought losses. Estimated at the lowest (and extralegal) rural wage ($0.35 per day), such labor may be valued at $7 million per year.

Adjustments made under governmental responsibility have included programs of meteorological, climatological, and crop-varietal research; weather modification, irrigation, and rural water-supply development; famine relief, food import and storage; and provision for migrants. Annual expenditure is estimated to be about $2.5 million, to make the total cost of private and public adjustments about $9.5 million. Annual average social cost (when valued in monetary terms) is about $18 million or 1.8 percent of gross national product — a $1.50 per capita cost in a nation in which per capita income is $75 per year. Thus, there is much that should encourage greater effort in loss reduction.

A heightened loss-reduction effort in Tanzania includes the development and adoption of drought-resistant crops, greater climatological sensitivity in planning regional specialization, more optimal policies of storage, pricing imports, and marketing of food grains, more rational collective use of range lands, and possibilities of improved drought husbandry through socialist organization.

From our observation, two trends seem to underline the need to explore a broadened range of public and on-farm policy. First, the Uhuru, or independence decade initiated in 1961, which largely shaped the working experience of policy makers, was favored with ample precipitation. Less well-watered decades are found in the precipitation records and can be expected to recur in the future. Second, there are many signs that the monetary social cost of drought events is increasing rapidly. The largest outlay for relief costs recorded to date was made in 1969, a year with 90 percent of normal precipitation, and was incurred because of increasing reliance on marketed grains, poor harvest in the three major areas of commercial grain production, and government extension of relief aid. In 1974–1975 over $100 million worth of food aid was imported. The full assessment of relief cost has yet to be made. Costs are high for farmers as well as for government. Farmers in nonfamine but drought-affected areas who do not receive relief must pay premium market prices for supplemental food up to 100 percent above normal. Finally, recurrent droughts seem to inspire costly settlement and development schemes.

The short-run prognosis is for escalating social costs in the transition period from subsistence to commercialized food grain production, when the relative security of the folk society is slowly forsaken and the protective institutions of market and government are still unformed. It is just during that period that the opportunities for cost and loss reduction from a comprehensive attack on drought may be most helpful in relieving the fear and misery of hunger in the land.

Australia

In this island of continental scale live only 13 million people, perhaps about 1 million of whom derive their livelihood directly from soil and grass (Figure 3.3). Numbers alone do not sufficiently account for the importance of agriculture to the economy. With 6.7 percent of the work force in rural pursuits, they account for 6 percent of the GNP and, more importantly for the economy, 50

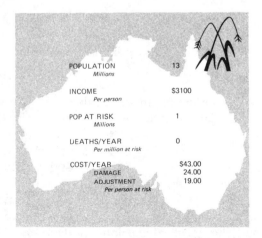

POPULATION		
Millions	13	
INCOME	$3100	
Per person		
POP AT RISK	1	
Millions		
DEATHS/YEAR	0	
Per million at risk		
COST/YEAR	$43.00	
DAMAGE	24.00	
ADJUSTMENT	19.00	
Per person at risk		

FIGURE 3.3. Australia: Drought.

percent of the foreign exchange. Four-fifths of the land is defined climatologically as arid or semi-arid. Meteorological drought measured by an accepted local standard (the lowest decile of historic rainfall) affected at least one-tenth of the continent in 32 of the last 85 years of record (Heathcote, 1969a). By this standard, serious drought affects substantial parts of the Australian continent 1 year in 3.

The effects of drought upon human activity appear somewhat different in the semi-arid pastoral plains of the eastern sectors than, for example, in the mixed wheat and sheep lands of the south and western sectors. In the sheep-raising grasslands of the east, agricultural occupance has fluctuated for over a century with sharp changes in stocking densities reflecting the variation in hazard events. This use Heathcote has described as "opportune," to distinguish it from official government intent in managing the grasslands:

> Drawing upon their increasing experience, the men on the land have met a fluctuating resource base by considerable mobility of effort both in time and space. They have stocked heavily in the good seasons because they knew that surplus feed would die off unused in the next drought; they have held onto as large an area as possible knowing that they have thereby reduced the risks of total failure in a localized drought; they have sent their stock all over the plains to find feed in widespread droughts; they have wherever possible acquired capital through partnerships, companies, and ultimately mortgages, to carry them over the bad season and to restock their empty paddocks after drought; and they have always welcomed the buyer, be he "fool" or wise man who would give them ready cash to walk off the land and leave it.
>
> . . . [I]f we credit the century of white occupation of those semi-arid lands with any success then the success owes . . . much to the timely exploitation of the resources by the men on the land. (Heathcote, 1969b)

In contrast to the opportunistic exploitation of the grasslands, the systematic occupation of the mallee (shrubby) lands of Victoria and South Australia exhibits a pattern of agricultural occupance emphasizing long-term stability, large capital, intensive technological innovation, and a growing and successful ability to cope with drought (Heathcote, 1969a, and our Chapter 4, which describes for a single farmer some of the techniques and managerial practices that have made this possible). The differential in strategies between the sheep lands and the wheat–sheep lands reflects in part the differences of soil, water, and landform, and in part the essential strategies of pastoralism and agriculture. In both instances success has come slowly, since the appraisal of land potential required experience of oscillations in rainfall and in the popular images and official optimism about the area encouraged by immigrant origins from more humid lands. After a century of effort, the degree of relative success is impressive, although it has not been acquired without cost. What is the magnitude of these costs?

Production loss over a period of 66 years has been estimated at $1.6 billion, or an average of $29 million per year (constant dollars). These include estimated losses from crop failures, livestock deaths, and associated economic deficits. Farmers and ranchers spend substantial sums to avoid drought losses. Among such expenditures should be included some portion of irrigation works and water supplies, a part of the machinery for water conservation, cultivation, and rapid replanting, some portion of the excess fodder stock inventory and similar expenditures. The cost of such on-farm adjustments and the extra production effort they entail can be estimated as a fraction of certain production costs and capital investment, or about $9 million annually. Added to these are the on-farm emergency costs of livestock transport, purchase of fodder, and the like.

For the million people directly at risk the damages run $20–$25 per year, and the adjustment costs are somewhat less. The benefits comprise a substantial part of the national income.

At the level of government expenditure there is a mix of official state and commonwealth drought-relief policies, a portion of monies expended for irrigation and water-supply development, emergency provision of water by rail or highway, subsidies for movement of surviving livestock, and the importation of relief fodder. The maintenance of an overseas credit reserve is in part a drought measure, and there is a substantial annual cost of drought related to research. A rough estimate of the average annual cost of all these items is $8–$12 million, or roughly $1.00 per person per year.

Other severe disaster effects in the past included the effect on settlement: land abandonment, reduction of the intensity of land use, depopulation, and failure of communications. For example, a program related to marginal wheat lands involving a reduction of use intensity from cropping to grazing in the 1920s and 1930s cost $4 million in subsidies alone.

Recent trends in official policy include the establishment in 1965 of a "drought watch" as an information service, the provision of "drought bonds" in 1969 as a form of pastoralist insurance, and continued study by the Australian Consultative Committee on

Drought. But these seem less important than the capital investment
for drought-prevention adjustments, the improved managerial skill,
and greater farm size and investment which in combination have
reduced this, the major continental natural hazard, to a manageable
burden.

FLOOD

Sri Lanka

Of the 12 million people in the island republic of Sri Lanka, about
two-thirds gain their living from agriculture and at least 25 percent
of those till land that is subject to recurrent flood (Figure 3.4). But
it is not only the peasant in his rice field who suffers from excess of
the stream flow that is essential to his crop. Increasingly, the city
dweller is susceptible to flood damage. According to studies by
Hewapathirane (1977) in both rural floodplain and city riverfront,
the risk enlarges in the face of heavy government expenditure for
levees, channel improvements, and upstream dams.

For the cultivator of rice in the wet western section of the island,
the rainfall is generally sufficient to harvest a crop. Some farmers
along the lower zones of floodplains have adapted their crop pattern
to annual overflow, and they would be disappointed if stream flow
were to fail. For others in higher zones the floods rising in

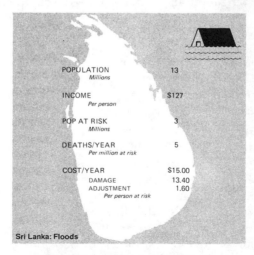

FIGURE 3.4. Sri Lanka: Floods.

October–December cause a decrease in crop yields and are to be feared rather than welcomed. In the dry eastern section, agriculture is almost wholly dependent upon irrigation, much of it from reservoirs large and small. Floods are a threat only when rare torrents surge over the usually dry watercourses or when the upstream dams fail, as in 1957. In that very wet year many small storage dams were ruptured by high flows, and downstream channel-improvement works were undercut.

The most highly built-up sections of the leading city of Colombo and of several other urban centers, notably Kalutara, Kalaniya, and Peliyagoda, are located in floodplains. These, with large industrial installations, are partially protected by levee systems and a few reservoirs upstream. At least 135,000 city dwellers were displaced in Colombo by a flood in 1947.

Continued enlargement of the number of people exposed to floods is due to two sets of factors. The rapid growth of total population since the early 1900s fostered increased densities in both rural village and city. In Colombo District the urban population grew from 145,000 in 1891 to more than a million in 1963, and the rural dwellers increased almost threefold. A large part of the growth edged farther down from the upland into floodplains that offered more land but at the price of recurrent inundation.

Such invasion was encouraged by the building of small levees along numerous streams during early periods of British management. The levees offered temporary protection against the floods they were designed to curb, and in that fashion they attracted still more people into these areas. When the levees break, however, the losses are tremendous. Moreover, the levees in some places prevent water from draining back to the confined streams, and thus cause prolonged flooding in theoretically protected areas.

To make matters worse, the channels of streams have been filling with sediment so that overbank flows have become more frequent. The causes of increased flood frequency include the progressive change in drainage-area land use through spread of plantation agriculture and forest clearing in the colonial period following British control in 1796, and the effects of channel improvement ultimately reduced the capacity of channels to carry stream flow. A decline in frequency of precipitation is believed to have begun in the 1920s. Channels silted, drainage was impaired, and much lowland began to go out of cultivation.

Rough estimates may be made of the amount of damage and of the net reduction in production that were caused by floods. Hewapathirane calculates that the reported loss for 1962–1967 amounted to about one-half of 1 percent of the gross national product — and this takes no account of other types of disruption such as injury to health or slowing of manufacturing. To the loss figures should be added public expenditure for flood protection. Because part of the new work is in multipurpose reservoirs in the dry zone, construction costs are difficult to allocate. As a minimum, the annual investment in works for flood control is on the order of $10 million. This is about 60 cents per capita for people at risk. The construction program began during colonial days and has been based on the advice and skills of British and, later, U.S. consultants.

Until the national attempts to curb stream flow by engineering works began in the nineteenth century, the prevailing adjustments were for urban dwellers to shun more hazardous floodplains and for the great body of peasants to adapt their daily practices to occasional inundation. For this purpose they used and still use a great variety of techniques, including raised house construction, careful choice of crop varieties, and emergency actions immediately before, during, and after a flood. The total of losses and adjustments amount in an average year to about $4 per capita. This amounts to 2.5 percent of the gross national product.

With the design and construction of larger engineering works, many fields and houses were given additional protection, while others were exposed to longer ponding of water and more frequent or more catastrophic overflow. The Meteorology Service instituted more precise flood forecasts for the Colombo area, and the government organized a more effective relief program for flood sufferers. Part of the agricultural loss is covered by a national system of crop insurance instituted in 1957. There is growing recognition in Sri Lanka that primary reliance upon engineering works to cope with flood losses is leading to larger losses while protecting selected areas. But emphasis in 1974 was on the engineering adjustment, and the short-range prospect was that it would continue.

The United States

After relying primarily upon a policy of heavy investment in flood control, the U.S. government experimented for a period of over four decades with a greater number of flood adjustments than any other industrial nation. In the course of its efforts to reduce flood loss by measures estimated to yield benefits greater than the costs, it went through a period of primary emphasis upon engineering works. By 1976, however, national flood policy was highly flexible, and the nation was beginning to work out a genuinely integrated and balanced strategy. About 10 percent of the population is exposed to some kind of flood threat, whether from small gully or major stream (Figure 3.5).

Until the Federal government became involved in flood control projects on the lower Mississippi and Sacramento rivers in 1917, the occupation of floodplains was a matter of personal or local government responsibility. Farmers and city folk moved within reach of floodwaters at their own risk and often in ignorance as they pushed across the continent. Where large tracts of land could not be farmed without elaborate drainage and protection works, the farmers banded together in local improvement districts and assessed themselves for the cost of building levees and drains, as in the Yazoo Delta.

Following the great flood of 1927 in the alluvial valley of the Mississippi, federal responsibility was extended to the lower river; and following the upper Ohio Basin floods of 1936, a national policy was adopted for building protection works "wherever the benefits to whomsoever they may accrue" exceeded the costs. A program of upstream watershed protection was begun at the same time.

Population growth and the provision of protection works contributed to further invasion of floodplains. By the late 1960s it was estimated that within the 5–10 percent of land area subject to flood there were all or parts of five thousand municipalities with population exceeding 2,500, and that about 60 million hectares of farm land were in floodplains. Much of the fresh invasion of floodplains takes place below detention dams in unprotected urban areas or in suburban development of farmland by people new to the

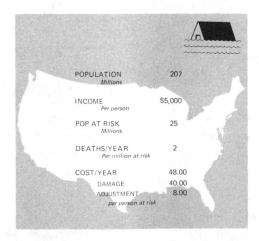

POPULATION *Millions*	207
INCOME *Per person*	$5,000
POP AT RISK *Millions*	25
DEATHS/YEAR *Per million at risk*	2
COST/YEAR	48.00
DAMAGE	40.00
ADJUSTMENT *per person at risk*	8.00

FIGURE 3.5. United States: Floods.

countryside. Direct flood losses up to the early 1970s were estimated at $1 billion or more annually, with the major part in urban areas. The worst natural disaster in the nation's history was the flooding caused by Tropical Storm Agnes in June 1972. The highest loss of life in one flood—238 deaths in all—occurred that same month in Rapid City, South Dakota.

Apparently the volume of average flood damage continued to rise after 1936, and per capita loss (now about $5) increased gradually. The average annual cost accounted for about 0.15 percent of the gross national product in 1936, and for three times that amount in 1972.

Federal expenditure for flood control also maintained a relatively even annual level (cumulative in 1975 about $9 billion). In addition, the federal government spent roughly $3 million annually on a flood-forecasting system, and supported a subsidized flood insurance system, relief and rehabilitation assistance to needy individuals and local governments, and research devoted to flood problems, chiefly hydrological. Voluntary contributions for flood victims were made through the American Red Cross. How much private-property owners pay annually to reduce the disruption of floods is not known with accuracy. Some floodproof their buildings in advance of flooding, by installing watertight walls and doors or elevating any fragile machinery they may own. Others pay premiums on more than $12 billion of insurance.

After 30 years of concentrating heavily upon engineering works to prevent flood losses, the federal government in 1966 turned slowly toward a mix of adjustments that gave primacy to protection but opened the way for other functions. Spurred by a task force report noting the continuing upward trend in flood losses (U.S., 1966), the government increased its emphasis on providing information on flood hazard and opportunities, curbed further federal building on floodplains, expanded its technical advice on floodproofing, and enlarged its flood-forecasting service. In offering flood insurance to occupants of vulnerable areas on condition that their communities adopt regulations against unwise development of floodplains, the government gave incentive to property owners to enjoy financial protection from flood losses without enticing others to augment the ranks of the exposed. It also stipulated that relief payments for people and public agencies in known hazard areas be conditional upon their having signed up for such insurance, and that federal mortgage insurance for houses in the floodplain be taken on insured property.

By 1976 the change in emphasis was beginning to be reflected in local decisions, but it was too early to judge what effect it was having on flood vulnerability. Translating policy into action at the local level has been slow and cumbersome.

TROPICAL CYCLONE

Bangladesh

The tropical cyclone threat in Bangladesh is concentrated in two seasons — April–May and October–December — when vigorous storms sweep up the Bay of Bengal. The storm surges arising from these have ravished the coast of the Bay of Bengal for many centuries. The first record of a devastating storm was written in 1584. Although the low-lying deltaic areas at the mouths of the Ganges-Brahmaputra (noted in Chapter 1) are the most vulnerable, the entire coastline of Bangladesh is subject to severe damage (see Figure 3.6). The seaward margins consist of a series of shifting islands (*chars*) formed and constantly altered by deposits of silt and sediment carried down by the many mouths of the rivers. Often these islands are bordered by shoals and sandbanks extending in places for miles, the monotony being broken occasionally, particularly in the western districts, by mangrove swamps where the influence of salt water is greater. About 15 percent of the nation's land area, containing some 10–15 million people in 1972, is subject to the direct effects of tropical cyclones.

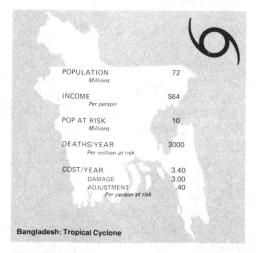

POPULATION *Millions*	72
INCOME *Per person*	$64
POP AT RISK *Millions*	10
DEATHS/YEAR *Per million at risk*	3000
COST/YEAR	3.40
DAMAGE	3.00
ADJUSTMENT *Per person at risk*	.40

Bangladesh: Tropical Cyclone

FIGURE 3.6. Bangladesh: Tropical cyclone.

Bangladesh is one of the more densely populated nations of the developing world. Its rapidly growing population—over 72 million people in 1974—constantly threatens to outgrow its food supply. Consequently, sons of farmers, many themselves sharecroppers, are faced with either feeding their families from progressively smaller plots of land or searching for new areas to settle. The press of hunger has forced villagers southward onto the fertile, flat deltaic plains; many of the *chars* subject to the full force of winds and storm surges of tropical cyclones were not settled until 60 years ago.

The deltaic plains are eminently suited to rice cultivation, and 80 percent of the workers are farmers, 10 percent fishermen, and the rest traders, clerks, day laborers, and professionals. A large body of migrant laborers come in to sow or harvest the crops. The people live in bamboo and straw houses, thatch-roofed and placed on mounds elevated sufficiently to keep them out of summer monsoon floods. Most houses are grouped in family compounds, and are often shaded by a few fruit and other trees (banana, date, coconut palm, and bamboo) growing close to them (Islam, 1974).

The farmer here has a very small margin of survival, so that he and his family are particularly vulnerable to any minor fluctuation in climate. In this setting the impact of a tropical cyclone is disastrous. In the cyclone of 12–13 November 1970 (described in Chapter 1), the massive surge striking the central coastal region brought death to more than 17 percent of the population of the affected area (Chen, 1973). Mortality was greatest among those who were too old, too young, or too weak to hold on to the trees that in many places offered almost the only place of refuge. On some offshore islands all the inhabitants were killed, and no one knows how many were lost of the migrant laborers accustomed to sleeping in the fields.

Though estimates of average annual damage of tropical cyclones in Bangladesh are rough at best, the 1970 figures, already cited, give a more precise idea of the impact of one storm. A study of that storm (Chen, 1973) indicates that 85 percent of the families in the area had their homes severely damaged or destroyed by the cyclone, and a total of 600,000 people were left completely homeless. Almost a million people were dependent for several months on outside relief food for survival. To return to agricultural self-sufficiency, it was estimated that 125,000 draft animals and 127,000 plows were needed. The monetary cost of the rehabilitation is difficult to estimate, but is reflected somewhat by a World Bank loan of $185 million for a reconstruction plan. The impact on the nutritional level of the country was to remain severe for at least five years. Approximately two-thirds of the fishing capacity of the coastal region was destroyed — and that in a land where 80 percent of the protein in human diet comes from fish (Frank and Hussain, 1971).

In the setting of a marginal and highly vulnerable existence, there are few resources for individuals to use in making adjustments to the threat of tropical cyclones. People do take active measures to protect themselves from the yearly storms by building raised-earth platforms in their houses, by tying down house roofs, and by planting trees to absorb wind and provide refuge in their tops (Islam, 1971). These measures afford little protection in an extreme storm. Moving away is scarcely an option, and farmers were observed resettling offshore islands that two months before had been swept bare of any trace of human habitation by the November 1970 storm (Chen, 1973).

The national government has concentrated on protective shore works, cross dams, and other structures (behind which newly formed land is used for agricultural purposes), a warning service, and protective plantings of trees along embankments. In the 1970 storm the protective works were breached, unable to withstand the waters that rose more than 7 meters above normal high tide. The tree plantings, useful in bank protection against small storms, were washed away in the larger one.

The government cyclone-warning services issued timely warnings of the storm, but they were not fully effective in 1970 because the social organization required to make them so proved inadequate. Radios, for example, are not available for all the population, and stations were off the air during the most critical hours of the night. In addition, although there were some community buildings and raised structures available for refuge, women accustomed to the seclusion of *purdah* were likely to resist evacuation to them.

The combination, since the 1970 storm, of intense economic pressure that drives the farmer and his family to settle low-lying lands, the government construction of protective works that offer a sense of security, and a higher level of relief and rehabilitation aid than in previous disasters is probably setting the stage for enormous destruction and loss of life in future severe storms unless much heavier attention is given to land-use patterns, building design, and warning systems.

The United States

In the United States, hurricanes (as tropical cyclones are known here) account for large loss of lives and property. The major components of hurricane hazard are storm surge, wind, inland flooding, coastal erosion, and tornadoes, with storm surge the principal cause of loss. About 30 million people are exposed to hurricane wind hazard, and some 6 million are directly subject to the storm surge arising from a storm, which is more likely to cause loss of life. About two hurricanes affect the East Coast per year, and from them winds of over 70 kilometers per hour may affect more than 5 million square kilometers of land, or nearly one-quarter of the nation's territory.

The losses caused by a hurricane are a function of the magnitude of the storm, the population, and the economic development of the area affected. In recent years there has been rapid development along the East Coast. The trend is for loss of life to decrease and for property damage to increase. The average annual property loss was estimated at $400 million for 1950–1970, and is believed to have increased to as much as $600 million by 1975 (Figure 3.7).

Individuals and families may suffer great losses, but these are somewhat compensated for by federal and local assistance. Communities are probably the social level hit hardest of all, for they suffer loss in public buildings and equipment such as schools and service facilities, and in addition have a diminished tax base after the disaster. They also receive heavy federal aid for rebuilding. Hurricanes cause significant damage to offshore installations such as oil rigs, fisheries, and boats of all kinds, this amounting to an average of $75 million per year.

Loss of life, amounting to some sixty fatalities a year since

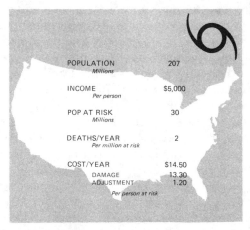

FIGURE 3.7. United States: Tropical cyclone.

1940, has decreased despite increasing population along the coast, but this downward trend may not continue much longer, for the growing urbanization of coastal areas means that some escape routes are becoming insufficient to take care of the population. One Florida county, for example, is exploring the possibility of vertical evacuation into sturdy high-rise buildings rather than attempting to get people out of the area endangered by the storm when evacuation routes may become clogged.

A recent response by the U.S. government to the hurricane hazard has been its experiments in modifying damage potential of the storm through seeding of the clouds. Since, however, the genesis and behavior of these storms is poorly understood, no one is entirely sure of the possible adverse effects of this procedure (such as reduction in rainfall or change in direction), and seeding operations might therefore cause increased damage.

In the past, protective shore works have been erected widely at a high cost, borne mostly by the federal government. Such works are now being criticized as inspiring a false sense of security, and also because they interfere with shoreline ecology and scenery values more recently significant on the American scene. Dunes and forest plantings have been used with some success in coping with minor storm effects, but these are unlikely to withstand a major storm.

The adjustment most directly related to saving lives has been the development of an effective warning system, with cooperation between federal and local officials, public and private sectors. Sugg (1967) estimates that such a warning system saves about $25 million during an average season, and perhaps as much as $100 million during a very active one. Overwarning may result in unnecessary protective or evacuation measures costing from $7 to $17 million. Each time a warning goes into effect in southern Florida, pro-grammed measures taken by corporations and businesses cost them about $2 million, regardless of the severity of the storm. Public response to warnings is not uniform, although up to a half-million persons were successfully evacuated in the case of Hurricane Carla, which went ashore near Corpus Christi, Texas, on 11 September 1961. Adequate explanation for this varied response is not available, but it may be related to prior experience with hurricanes.

Land-use management and the modification of building re-quirements are used as adjustments for lessening the deleterious effects of hurricanes. The potential payoffs from coastal and land management are reflected in the value of property in storm-surge hazardous areas, estimated at $25 billion by Friedman (1971). Other measures available to individuals to reduce hurricane damage include protection against flooding, and windproofing/emergency measures such as tying down loose materials and shuttering windows.

Despite the considerable body of knowledge available to the American people about possible adjustments to hurricanes, adoption of such measures is not widespread. For example, in the vulnerable coastal area of Miami, Florida, a leading shutter and awning company estimates that only 20 percent of the residents take the protective measures open to them (Sugg, 1967). Possibilities of vertical evacuation have only begun to be canvassed systematically. The trend of increasing damage as more and more people move into expensive structures on the coastal areas and as mobile homes proliferate may be expected to continue. The downward trend in loss of life may also be reversed if, as more people move in, no means are provided for evacuating them in an emergency.

AIR POLLUTION

In global perspective, air pollution is still very much of natural origin. The suspended particulate matter of the atmosphere from a great volcanic eruption dwarfs the input from the man-made environment. In local perspective, however, air pollution as a hazard is very much the product of the industrial revolution, bringing together new and continuing sources of pollution — dust, smoke, gas, automobiles — with great concentrations of population. Comparison is made between a nation where the industrial revolution began and a developing nation well exemplified by the Mexican experience.

Mexico

With a population of 53 million people (approximating that of the United Kingdom), Mexico has but one-third of U.K. income, one-tenth of the motor vehicles, one-fifth of the industry; yet at least 9 million of its people are exposed to a continuous and serious air-pollution hazard. Indeed, the air pollution of Mexico City is comparable to, and by some measurements exceeds that of, other great North American cities (see Figure 3.8).

With half its labor force and income still agricultural, Mexico exemplifies the paradox of the developing nation: coping with the recurrent hazards of flood and drought of an agricultural society and at the same time dealing with hazards created by modern

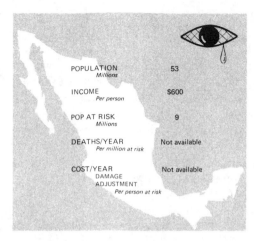

FIGURE 3.8. Mexico: Air pollution.

technology. In another sense, Mexico is atypical of developing countries. Its rate of industrial development is high and sustained, and its great metropolis of Mexico City is peculiarly vulnerable topographically and climatologically to air pollution. Nonetheless, as Mexico moves into enforcement of its recently enacted legislation to prevent unlimited environmental pollution, its experience is enlightening.

Atmospheric pollution from windblown dust from the nearby desiccated lake beds, or from erosion of the surrounding mountains, has been a common feature of twentieth-century settlement in the basin in which Mexico City's 7 million inhabitants now dwell. The earliest reports of serious hazard from man-made pollutants originated in Poza Rica, Veracruz State, where, as early as 1922 and again in 1950, oil-refining accidents leading to gas emissions killed or seriously sickened workers and residents of the neighborhood. In the 1950s university-based studies began to document the high levels of serious pollutants in the Mexico City air and to relate these to the enclosed topography of the basin, the prevailing winds from the industrial northern section of the city, and the frequent temperature inversions that combine to retain the accumulated pollutants in concentrated form.

It was not until 1957 that official agencies other than a university group began to monitor the more important pollution indicators with a grid now extending to eleven points in Mexico City. Monitoring was planned for two other larger cities — Monterrey and Guadalajara — where serious pollution is believed to occur. Even though pollution levels are high, particularly for suspended particulate matter and SO_2, a debate exists as to whether the monitoring methods used by the governmental agency in cooperation with the World Health Organization are appropriate, whether they underestimate the pollution, and whether a recent one-year drop in some indicators is indicative of progress in control.

Control began in 1971 with a federal law to "Prevent and Control Environmental Pollution" of air, water, and soil, followed later that year by promulgation of "Regulations to Prevent and Control Air Pollution Caused by Dust and Smoke." The Subsecretaria de Mejoramieto del Ambiente (SMA), an agency for "improvement of the environment," was formed in the Department of Health and Assistance to coordinate and oversee all environmental protection activities to be carried out under the 1971 law.

The regulations are designed to control, in two stages, the emissions from stationary sources. The first stage allows industries to assess the amount and type of emissions they produce and to elaborate a plan whereby these emissions will be abated in accordance with federal guidelines. Factories are then to have antipollution equipment installed. This stage ended in May 1974; after that date SMA was to police the industries and if necessary file reports on violations. The respective Chambers of Commerce and the SMA claim considerable compliance. The cost of all types of pollution control has been estimated by industries to exceed $130 million by the end of 1973, and to rise to $350 million by 1976.

Although control of vehicle pollution is still in the project stage, it is likely that in the future some type of emission control device will be required. Another possibility under consideration is to reduce the high level of sulfur and lead in Mexican gasoline—levels in Mexico City are double that of New York and Los Angeles.

So far there has been no attempt to calculate the annual damage either to property or health, and officials are cautious about recognizing a specific health hazard. Nonetheless, the first epidemiological study, completed in 1974, correlated mortality rates due to respiratory illnesses in two high- and two low-polluted areas of Mexico City (July 1974). Preliminary findings indicate that higher levels of pollution are linked to increases in mortality among susceptible sectors of the population (children, the elderly, the chronically ill), but the results, as in such studies in general, are inconclusive.

Mexico moved rapidly in recognizing a long-standing problem and proceeding toward its control. This was partly in response to international efforts and publicity surrounding general environmental issues at the Stockholm U.N. Conference on the Human Environment (1972), and partly as a result of the general welfare thrust of its then current leadership. The government instigated a national policy and drew industry closely into its planning and regulatory activity. At present, however, efforts are directed solely toward the 9 million inhabitants of the three major cities. It is premature to evaluate the efficacy of these efforts, especially in a country where the drive for industrial development is not likely to lessen.

The United Kingdom

In 1863, the year the Alkali Etc. Works Regulation Act was passed to control emissions from factories manufacturing alkalies, the population of the United Kingdom was 25 million. Today, with a population of more than twice that number, this nation, among the first to undertake the industrial revolution, still suffers a high incidence of health injury, agricultural damage, and corrosion and clean-up costs. A substantial effort to control industrial effluents, a dramatic improvement in smoke levels, and a hesitant effort to cope with such newly recognized contaminants as heavy metals are salient elements in the social response to air pollution in the United Kingdom (see Figure 3.9).

Attention was drawn in a dramatic way to the air pollution problem by the famous London smog episode of December 1952.

FIGURE 3.9. United Kingdom: Air pollution.

The city was virtually paralyzed for four days, and a number of "excess deaths," estimated by some observers at 4,700, occurred. This event was the catalyst. Whereas pollution had been regarded as a highly localized problem affecting those in the vicinity of certain industrial establishments, and as a chronic problem continuing on a low level and having certain long-term effects, there was now dramatic evidence that millions of people could be affected over a short period of time. Air pollution disasters of major proportions were a threat to be reckoned with.

This smog event led to an enquiry, a committee report known as the Beaver Report, and subsequently the Clean Air Act of 1956. The focus of concern was switched away from industry, and the domestic coal fire—a traditional feature of the British hearth-centered culture—was identified as the prime source of the trouble. The London smog was attributed largely to a combination of smoke from domestic sources and of sulfur dioxide from both domestic and industrial sources.

The Beaver Report concluded that "air pollution was in 1975 costing the nation about $625 million a year in terms only of losses that can be given a monetary value." The major item was identified as health effects, especially bronchitis. The death rate from bronchitis, pneumonia, and other respiratory diseases was shown to be much higher in the larger urban areas than in the rest of the country, and higher in Britain than in any other country. Air pollution as a hazard was thus viewed largely in terms of smoke and sulfur dioxide, and was seen to have chronic long-term effects upon urban industrial populations, as well as having disaster potential in the event of a recurrence of a serious smog episode.

The Clean Air Act of 1956 (and as subsequently amended) set out to get rid of the smoke. The prime target for attack was the "black areas." Based on local authority boundaries, these were defined in general terms as the main industrial areas, and the areas of dense populations, where frequency of fog was also a consideration. The population of the "black areas" is approximately 25 million, or about half the total population. These are the prime "people at risk"—although to some degree everyone is exposed.

Under the act local authorities were empowered to define smoke-control areas, where the emission of dark smoke would be an offense. This required the conversion of many domestic heating systems from coal fires to other forms. The minimal change is to convert the fireplace so that it is capable of burning manufactured "smokeless" fuels. The property owner can claim 70 percent of the "reasonable cost" of the necessary conversion from the local authority, which in turn can claim from the central government.

The earliest designation of smoke-control areas has been in London, where the results seem to have been spectacular. Hours of winter sunshine in central London have been increased by over 50 percent, delays due to poor visibility at Heathrow airport have declined, and birds and flowers are reported to have reinvaded areas of the city where they have not been seen for generations. In the wake of this improvement, many historical and monumental buildings are being cleaned and are expected to stay clean for a longer time. The same rate of adoption has not occurred in some northern industrial cities nor in the urbanized areas on the coalfields.

The experience of the first 16 years of the Clean Air Act seems to have led to a situation in which a few spectacular results have been achieved, with the effect of generating a low level of public concern in which there is little recognition of individual role or responsibility. The battle against smoke has been won in the places where victory could be most readily obtained; and while the campaign has slowed considerably, further declines in smoke emission from domestic sources may be confidently expected even in the coalfield areas. The signs are that the threat of further smog disasters has been largely averted and that the chronic low-level problem of smoke will continue to improve slowly in line with general economic and technical trends.

The possibility that new air-pollution hazards may come to replace the old is, however, a subject of considerable research and active surveillance. Lead concentrations in the atmosphere and soil near places of heavy automobile traffic, and the leakage of lead and other heavy metals from industrial establishments, are newly recognized threats. Increased concentrations of automobiles, plus the higher incidence of sunshine resulting from the reduction of smoke, also create a potential for the development of "photochemical" smog. This latter possibility is currently the subject of vigorous denials on the ground that warmer temperatures and more intense sunlight are needed for "California-type" pollution. Nevertheless, some scientific evidence has been reported for the occurrence of photochemical smog in the south of England.

It is possible, but not established, that costs from air pollution in the United Kingdom have been rising as the level of expenditure on air-pollution control measures has risen. Certainly more recent estimates of cost have been higher than those made in 1954 by the Beaver Committee. Then $625 million was the figure aimed at, and now people are inclined to talk on the order of $2 billion. This is due at least in part to improved methods of assessing damage and to inclusion of a wider set of damage data in the calculations.

While overall damage trends remain unclear, it is apparent that improvements in one direction (smoke control) have revealed, and perhaps exacerbated, other problems. A result, in part, of adjustments carried out under the Clean Air Act, it also reflects the effect of long-term technological change. With a rising standard of living, and new fuel supplies of oil and natural gas from the North Sea, smoke in Britain's atmosphere would have declined anyway, and along with that a decline in the Englishman's disease of bronchitis and other respiratory infections was to be expected. New technology and economic growth have thus served to alleviate one well-known and highly visible environmental hazard, and have substituted an uncertain possibility of other less visible, less well-known, and less understood hazards (Burton et al., 1974).

HAZARD AND THE NATIONAL EXPERIENCE

The rains fall infrequently on the uplands of Tanzania or the mallee lands of South Australia; the flood waters of the Mississippi rise over days or weeks; the mountain streams of the west of Sri Lanka, in hours. The 80-knot-per-hour winds that sweep up the Bay of Bengal or across the Florida coast can be observed on weather charts or from satellite photographs for hours or even days before reaching their rendezvous with the dwellers on the shore. The effluent of human settlement and auto exhaust in London or Mexico City is always in the atmosphere, and increases or declines with the changing pace of economic activity and the changing seasonal and daily patterns of weather. In certain fundamental ways, human response to the environment as hazard is constrained by characteristics of the natural events, characteristics that seem to transcend differences in culture and material wealth.

The events studied in these brief surveys of national experience are reported in greater detail elsewhere (Visvader and Burton, 1974; Heathcote, 1974; Islam, 1971; Hewapathirane, 1977) and cover only a portion of the spectrum of characteristics that can be described as pervasive (frequent, slow of onset, with low-impact energy) or intensive (rare, sudden, with high-impact energy). Agricultural droughts are pervasive events; floods and tropical cyclones combine aspects of both pervasive and intensive events. Examples of intensive earthquakes or tornadoes are not included in the survey of national experience because of lack of comparative studies. Until more detailed investigations can be completed, much of the judgment must remain tentative.

The varied combinations of frequency, magnitude, and onset that characterize different events vary widely in the social costs they impose. Floods are costliest in damage and financial loss; cyclones, in loss of fife. That floods are the most damaging economically is probably due to the generally high utility of floodplains for transport and agriculture, combined with the middle range of frequency that characterizes the natural event. In terms of social impact, hazards may be thought of as varying along two dimensions: resource need and event frequency. Resource need depends upon the national context: floodplain land is more important to the welfare of Sri Lanka than to that of the United States. Drought is less crippling to the national welfare of Australia than to that of Tanzania.

Pervasive events are frequent and provide for a built-in warning; the progressive onset of drought is apparent to all farmers. Various degrees of warning are possible for floods and cyclones, but only short-term warnings are possible for tornadoes. Earthquakes may be in a transition stage with regard to warnings. Similarly, human capacity for dealing with high-energy hazards is more limited than it is for low-energy ones. All these factors appear to be reflected in the balance of human adjustment between suffering loss and coping with the events. When stated as the ratio of damage loss to adjustment expenditure, this ratio varies in national studies as between approximately 1:1 for drought and 5–10:1 for floods and tropical cyclones. This reflects the higher frequency and progressive onset of drought, and the comparative rarity, sudden onset, and high energy of floods and cyclones. Thus, while the per capita hazard costs are surprisingly close for each set of countries (developing and industrialized), the mix between suffering loss and coping with it is structured by the characteristics of the hazard event.

DEVELOPMENT AND THE NATIONAL EXPERIENCE

The costs incurred by people in the human use of the earth are continuous, and no nation escapes them. The basic trend of a high death rate in developing countries and a high and rising damage rate in industrial countries is substantiated by national studies. Within that basic trend, some experience is still exceptional. The Bangladesh cyclone of November 1970 was the worst natural disaster for at least several centuries in loss-of-life terms. Its largest predecessor was the Shensi earthquake of 1556, with more than 800,000 deaths. In 1976, reports were received of another Chinese earthquake, at Tanshan, which may have taken a death toll reaching up to 750,000.

Industrial nations exhibit a measure of success in different ways of coping with their hazard problems. The Australian experience (corroborated by experience in North America) suggests that highly skilled and mechanized large-scale agriculture and livestock enterprises can absorb enough drought that its cost can be carried by the enterprise without excessive disruption to the flow of either income or product, and with limited material assistance. This success of "technique"—the combination of skilled methods and labor-saving machinery—parallels that of the United Kingdom in dealing with smoke. In both countries, ongoing economic and technical trends lead in the direction of reducing loss potential. The Clean Air Act in the United Kingdom is commonly described as "swimming with the tide" of industrial technology. In Australia the trend to larger-sized holdings and less labor-intensive techniques serves to spread risk and increase flexibility. The new U.S. policy toward floods is hopeful and more comprehensive than either U.K. air pollution or Australian drought policies, but has not yet achieved its purpose of stemming the rising costs of hazard.

The costs of hazard—the loss from property damage and expenditure on adjustment—must be seen relative to the resources of the society. In that perspective the position of the developing countries is more serious. Figure 3.10 shows the cost of hazard per capita and as a proportion of GNP. In all cases the costs of hazard for developing societies expressed as a proportion of income available are at least ten times those of industrial nations.

In the developing countries the economic impact of loss from natural hazards is more serious than is commonly recognized, and it seems likely to remain a disproportionate drain on industrial wealth unless development strategies are designed with the natural hazard problem in mind. But to document the lower capacity of lesser-developed nations to deal with external natural events is only to add to an already large catalogue of their problems. The failure of industrial nations to stem the rising tide of loss is a salutary lesson, suggesting that the way to reduce the vulnerability of less-developed nations does not lie in emulating the experience of industrial nations. Where this has been done—in reliance on engineering works in Bangladesh and Sri Lanka, in providing extensive central government relief in Tanzania, or in improving meteorological skill

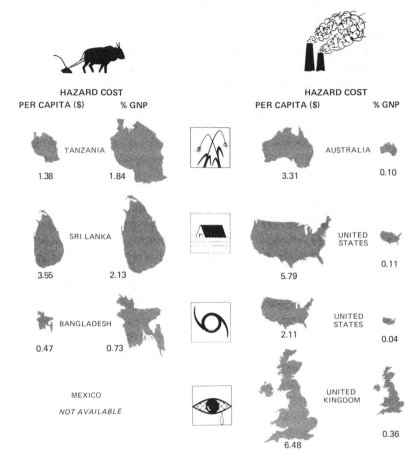

HAZARD COST
PER CAPITA ($) % GNP

HAZARD COST
PER CAPITA ($) % GNP

TANZANIA
1.38 1.84

AUSTRALIA
3.31 0.10

SRI LANKA
3.55 2.13

UNITED STATES
5.79 0.11

BANGLADESH
0.47 0.73

UNITED STATES
2.11 0.04

MEXICO
NOT AVAILABLE

UNITED KINGDOM
6.48 0.36

FIGURE 3.10. National studies: Hazard costs and national income. Hazard cost is the sum of damage loss and costs of adjustment expressed here as annual average per capita costs and as a percentage of gross national product. The dollar value of per capita loss varies much less than GNP. The loss of GNP is about 20 times greater in developing countries.

without the social capability of warning in Bangladesh — the results have been costly. In a comparative context, this is not surprising. Reliance on engineering works has been costly in the United States, and response to warning still fails tragically, as in Rapid City, South Dakota, or is poorly formed vis-à-vis air pollution in the United Kingdom.

There are alternative paths that may effectively preserve the knowledge and wisdom of the traditional society while adapting it to the needs and capabilities of the transitional society and blending it with new and established technologies. Any particular path is carefully structured by the special characteristics of the hazard of concern and the unique vulnerability of the society in question. While we assert the existence of alternatives, we cannot point to well-trodden paths: these await discovery and development by processes of individual, collective, national, and international choice.

Individual Choice

The farmer in the Usambara mountains of Tanzania and the farmer in the southern Australian plains contemplating a slowly developing drought face similar decisions as to whether to remain in the hazard area and what they will do in the event of continuing shortage. Like other individuals who knowingly expose themselves to a natural hazard, they make some kind of appraisal of the prospect that drought will continue. They canvass a number of the possible actions to be taken in dealing with the threatening environment. Each thinks about the consequences of taking those actions to himself and his family. Finally, they choose in one way or another what, if anything, they will do.

One farmer may plant a "catch" crop to try again for at least a modicum yield. The other may quietly sit it out. As outlined in Chapter 2, they can do many things, and they can decide to do nothing. It is this process of individual choice that is at the base of much of the action of people in dealing with extreme natural events. The invasion of Char Jabbar by peasants is triggered by government's decision to protect the coast, but the actual movement is an aggregation of individual decisions. The decision of the Wilkes-Barre citizens to remain in their homes after the flood receded is reflected in a public program, but it is at root a summation of family actions.

The land behind Baghdad, like the ruins of Mohenjo-Daro, gives mute testimony of failure to survive in places of hazard. The density of settlement amid poverty in northeast Brazil, as in the Ganges–Brahmaputra delta, testifies to survival without prosperity. With regard to the less chronic or less cataclysmic events, why is it that people appear at times to seek less than the best for themselves, to take unnecessary risks, to be ignorant of serious consequences, or to seem remarkably foolish in a hindsight view of their personal history?

If it were known precisely why people select some information and ignore other information, much social behavior could be explained. For example, risk taking by residents on the seismically active San Andreas fault zone in California is often dismissed with facile generalizations. Residents of the fault zone are branded simply as greedy or stupid or shortsighted. Their actions are then explained or forecast in similarly simplistic terms. These appraisals come easy, are applied readily to public solutions, and are usually misleading. They rarely have predictive power; more complicated answers must be sought.

For an understanding of how choices are made, it would be helpful to have a detailed theory of individual decision processes. But, as indicated in Chapter 2, no satisfactory model is yet available to illuminate the behavior of the Bengal fisherman or of a Tanzanian or Australian farmer. Lacking that, we can develop a rough model of bounded rationality to review what is known and to outline relations that still are speculative.

HOW DO PEOPLE CHOOSE ADJUSTMENTS?

The significance of the choice process and the difficulties of probing it are illustrated by two men of the land who face drought. One is in an area just beginning to undergo modernization in a semi-arid section of Tanzania. The other makes use of highly refined agricultural technology in growing wheat and wool in southern Australia.

Hamisi Juma

Juma lives on the southeast slopes of the western Usambara Mountains. Rising from about 400 meters to 1,800 meters in elevation as a small, compact, uplifted block in northeast Tanzania, the Usambaras are densely populated by Shambala-speaking people. On very steep slopes of up to 45 degrees, some terraced, Juma attempts to support a household of six persons. He cultivates 2 hectares of land in five separate plots planted primarily in maize, beans, and cassava. The family food supply is supplemented by bananas grown in moist gullies. Since he is far from more affluent centers, Juma has few opportunities for cash crops. His present mainstay is the spice cardamom.

Compared to that of neighbors farming higher up the slope, the moisture available to Juma is moderate at best and highly seasonal.

Average rainfall is somewhere around 850 millimeters, spread over a usable season (where rainfall is greater than one-half the potential evapotranspiration) of five months. It comes partly in November–December and lags in January–February; the greatest amounts are available in March, April, and May.

When interviewed, Juma is eager to talk about drought. The previous cultivation year (1969–1970) had been a harsh one; the out-come of the then current year (1970–1971) was still uncertain. At the age of 41, he recalls two other droughts besides the most recent one, and estimates that three years out of ten were bad ones for him and his family. When asked what he does about droughts, he readily explains different actions. In addition to prayer, these include increased planting of cassava (a crop with low moisture requirements and high storage potential in the ground), planting sweet potatoes in swampy ground, and stopping maize planting in order to save seed and labor. He will seek wage-work to buy food, although he is not hopeful of finding it and could ask for help from family and friends. Upon further discussion and in the light of information from an agricultural worker, Juma affirms some additional action: he attempts to plant early, and with enough rain, he staggers his planting and his plots, he weeds more in order to reduce plant competition for moisture, and he asks for help from government representatives.

Whether on his own initiative or that of his community and government, Hamisi Juma employs about one-half the potential set of adjustments available to him as conscious drought-related action. These adjustments differ in their function, in their time of initiation, in the responsibility for initiating them, and in their potential cost and benefits. They can be compared in these respects with those available to Bill Bauer, whose 800 hectares of bought and leased sheep and wheat land in south Australia dwarfs the small holdings of Hamisi Juma. The drought adjustments available to both men are not dissimilar in function, although there are substantial differences in potential costs, benefits, and timing.

Bill Bauer

The Bauer property is in the mallee lands of southern Australia, an undulating plain approximately 80 meters above sea level, with long, low sand ridges (about ¼ kilometer wide) separated by broad, loamy flats 1–2 kilometers wide, oriented east–west. Originally this country was covered by dense eucalyptus scrub some 1–15 meters high, with a thin grass cover beneath. Since World War I more than

80 percent has been cleared for wheat crops and improved sheep pastures. Bauer's family is now there; his eldest son has a farm of his own 600 kilometers away in New South Wales. Bauer's place is mainly in one piece and freehold, but with two separated blocks (5 kilometers and 10 kilometers distant) rented from neighbors. Fields are 50–100 hectares apiece, the rectangular grid of wire fence lines looping indifferently over sand ridge and flat. Each year in Autumn (March to May) one-third of his land is fertilized and sown to wheat, one-sixth lies fallow, conserving plant nutrients and moisture for next season's crop, and half of the remainder is sown to lucerne or ryegrass, and the other half left in wheat stubble and native pastures, both as fodder for his thousand head of sheep.

Bauer grows none of his own food except for eggs from the hens kept by his wife and an occasional mutton from a slaughtered sheep. Each year he sells his wheat crop to the Australian Wheat Board at a fixed price and gambles that his wool clip and culled surplus sheep will fetch a good price at the open city auctions in Adelaide. His wheat money provides the necessities of food and clothing, the canned vegetables, preserved meats, frozen fish, and fresh bread (to be frozen in the deep freezer he bought when the rural electricity grid came through ten years ago) from the local town store opposite the wheat silo and the railway siding some 40 kilometers (a half-hour's drive) away. The wool income provides the extras: the electric appliances to ease his wife's housekeeping chores, the next bigger combine harvester, and the occasional new car or utility truck.

Bauer lives with drought. Average rainfall is 345 millimeters, and the chance of drought is for four years in every ten. If sufficient rain for seeding has not fallen by 1 June, he knows that he is in for a dry year. Some of his neighbors sow at the same time each year regardless of conditions, but he prefers to wait for the "opening rains" of April and May, reasoning that if they do not come he has saved his seed for later. On occasion he has sown as late as October, and then only to provide some feed for his starving sheep when summer starts in December.

At age 50 he has outstayed almost half the original settlers he can remember from boyhood, and has survived four major droughts and innumerable dry spells (he started, as did the others, with 400 hectares in 1944). His survival, he claims, comes from hard-earned experience and ability to invest spare capital in more land rather than in the latest car. He is not much given to prayer, and was scornful of government cloud-seeding experiments during the 1965 droughts. His reaction to drought was to sit it out on his previous year's cash surplus and, with the help of his stock agent or bank manager, to keep up the mortgage payments on his land and the latest combine. As soon as any break in the drought showed, he put in a crop and hoped for the best; otherwise he held tight, expecting to sell off his surplus breeding ewes before the bottom dropped out of the livestock market. When questioned further, he admits the value of the latest government-sponsored wheat breeds, admits, too, that rust is a problem in wet years (having tried to plant some of each variety as they became available), but feels that if drought does not extend beyond two years he can ride out the loss of crop, the reduced income from livestock, and the sanded-up fences and livestock drinking troughs.

Mechanization, he feels, has helped considerably. Comparing his motorized equipment with horse teams during the 1944–1945 drought, he claims that tractors and self-propelled harvesters had enabled ground to be prepared 10 times faster after opening rains, crops to be reaped faster in difficult harvest weather, and blowing sand ridges to be sown down to ryegrasses and held, as well as providing grazing for his sheep. He no longer needs to plant or buy oats for horses, and, if necessary, tractors can be worked all night with his son's help.

He does not approve in principle of official drought relief as currently offered (mainly local road works, hire of farmers' trucks, and/or actual employment on road maintenance or construction), feeling that those neighbors who applied for it were inherently bad managers anyhow. Government, he claims, could do more by reducing the costs of farm production, especially freight rates to and from markets. Too many folk, particularly in the news media, make too much ballyhoo about drought; he would prefer greater concern for the rising costs of farm production. He feels that he can cope with drought, but of his ability to cope with costs he is much less confident. In 1971 he was concerned more about the government's quotas on wheat production (since removed because of world wheat shortages) than about the current season's dryness.

How do Hamisi Juma and Bill Bauer arrive at their particular allocations of time, land, and capital in the face of ever-threatening drought? To what extent can observations of their choices be generalized for other people and other hazards? Do their methods of appraising the hazard and alternative actions, and their selection of adjustments to drought, illustrate the choices made by others facing drought or flood or cyclone?

ELEMENTS IN THE CHOICE PROCESS

People who have been observed coping with risk and uncertainty in the environment appear to take account of likely economic out-comes, as is suggested in the case of the Bengal fisherman, but the resulting behavior rarely conforms to what should be optimum by their standards of utility. They do not seem to maximize consistently the expected gains of their actions. It is easy to point out some of the difficulties encountered in canvassing the elements, as diagrammed in Figure 4.1. They cannot appraise the magnitude and frequency of extreme events — states of nature — with accuracy. Hydrologists and climatologists themselves have trouble estimating a great drought or recurrence of strong winds.

People are rarely aware of all the alternatives open to them. They differ greatly in the way they judge the consequences of particular actions even on the rare occasions when the physical outcomes are known accurately. The comparison of many different consequences is a highly complex operation for a decision analyst armed with precise data and a computer, let alone for a farmer choosing a crop as the rainy season approaches. A bounded

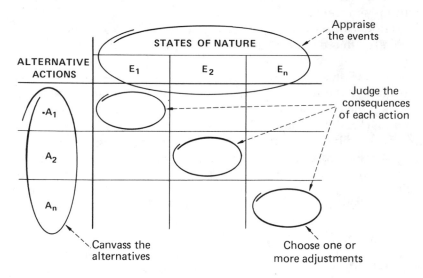

FIGURE 4.1. Major elements in decision behavior. Underlying the payoff matrix (see Figure 2.8) is an ordering of the major elements in decision behavior. To determine the states of nature (columns) requires an appraisal of natural events, and a canvass of alternatives is needed to specify the available actions (rows). The consequence of each action under a given state of nature (cells) needs to be evaluated as a basis for one or more choices.

rationality model that comes close to outlining how people in fact behave is likely to be far more complex. The most elaborate attempt to do so is that by Kates (1971), but in a model that has not been tested in a rigorous fashion. It can be simplified by extracting four elements that are noted frequently in field observations and by defining the points at which particular factors in some cases are known to play a significant part in the decision. An elementary scheme (shown in Figure 4.2), drawing from many previous analyses of decision making (Slovic et al., 1974), suggests that the individual (1) appraises the probability and magnitude of extreme events, (2) canvasses the range of possible alternative actions, and after (3) evaluating the consequences of selected actions, (4) chooses one or a combination of actions.

Both field observations and psychological laboratory work would indicate that the appraisal of extreme events and the canvass of alternatives are interdependent, and that people have difficulty evaluating more than a very small number of alternatives at one

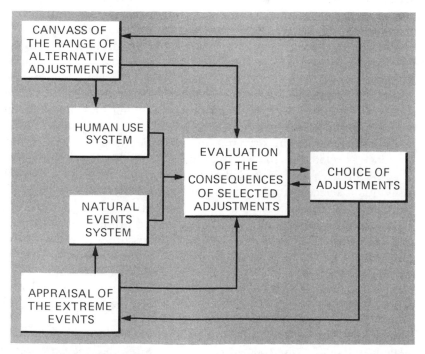

FIGURE 4.2. Simplified choice model. The major elements of the decision model (see Figure 4.1), with the interactive model of nature and man (see Figure 2.1), provides the basis for the simplified choice model used to present the findings of field and laboratory observations of individual choice.

time. There also is reason to believe that people tend to deal with alternatives in an ordered sequence rather than simultaneously (Kunreuther, 1974). It is essentially a process of information flow and interpretation, one that may work in a rudimentary fashion, taking many shortcuts, or may be intricate and convoluted. In any circumstance the individual selects and weighs information from natural systems to arrive at judgments as to the "state of nature" and as to which state is desirable and practicable of attainment.

It seems likely that the process of choice does not begin until after a first threshold of awareness of actual or anticipated loss is reached. But the limit may vary greatly among people: some people ignore or fail to perceive environmental risks that would stir others to consider alternatives. To exceed that limit induces the person to think about an extreme event and of ways of dealing with it. Up until that time, although damage may be suffered from the event, it is absorbed in some fashion as a part of the use system. When the second, or loss, threshold is crossed, need arises for purposeful action to modify or remedy the loss. Short of moving out of the place, the individual then searches for some finite number of alternatives involving changes in resource use or technology.

When the reasons for migrating become sufficiently compelling, the third, or use-location, threshold is crossed, a new location is found, and the same process begins again whenever damage occurs or threatens. There is no need to assume that people go through precisely the same steps in approaching or traversing the three thresholds. In many instances, however, they appear to follow a certain sequence in arriving at an ordered choice, and that sequence is related in part to the character of the thresholds.

"Ordered choice" means that type of choice process by which individuals select among alternatives by a series of two or more judgments among successively restricted options. Some economists with a certain disdain for dictionaries as the product of lexicographers call this a "lexicographic" approach. It assumes that people can compare only a relatively small number of alternative actions at any one time and that they apply criteria of evaluation of the consequences in a rough order of importance. The housewife at a display case to buy foodstuff does not consider all possible foods at the same time; she selects from one category at a time.

In East African households in which the housewife is obliged to walk each day to draw domestic water, she often has a large number of possible sources — springs, wells, streams — within walking distance. Even in the dry season she may have more than one alternative. Only rarely is she able to name more than seven as meriting consideration. The woman responsible for drawing water does not necessarily go to the nearest source. She often seems to rate the available sources as good or bad in terms of whether or not she (rather than the water expert) considers it a hazard to health. From among those sources judged "good," she then tends to select the one having the lowest economic cost in terms of caloric expenditure in carrying the load home. Even then, she may avoid the most economic source because she considers that other factors (such as irritating encounters with neighbors) may outweigh the economic benefits.

As outlined in Figure 4.3, the process is believed to be one of ordered choice, moving from a primary criterion of quality to a secondary criterion of economic cost, and occasionally from there to consideration of social effects. To the extent that this is, in fact, the procedure for people selecting or rejecting adjustments to natural hazard, it suggests that they move by steps that are related to critical thresholds defined by their particular criteria for evaluation.

COMPARATIVE STUDY SITES

The evidence to support or challenge these propositions and speculations about individual behavior in the face of capricious nature comes partly from scanning the experience of national groups, partly from relevant laboratory experiments, and partly from detailed observations in field situations. Much of the field work was interdisciplinary, and all of it was cross-cultural. As noted earlier, such work encounters severe obstacles, of both method and communication, which make for ambiguity of concepts and findings.

The history of national government efforts to cope with hazards, however diverse, has rarely been subject to sufficiently searching analysis to permit general conclusions as to the ways in which individual and collective choices are made. Careful comparison of field observations is largely lacking. Moreover, no laboratory trials of behavior in confrontation with natural hazards were undertaken until 1967, when a series of investigations was made in Toronto of measures of risk perception and behavior. In 1973, experiments with purchase of flood insurance and earthquake insurance were launched by Kunreuther (1977). Still, a number of carefully controlled experiments on risk-taking behavior, involving observations in the plush gaming rooms of Las Vegas, Nevada, in the deeply carpeted offices of stockbrokers, and in the leaner quarters of nursing school students, throw a little light on the

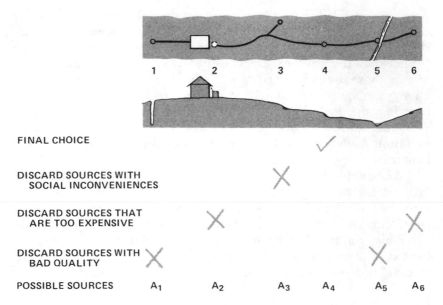

FIGURE 4.3. An ordered choice among possibly hazardous water sources in East Africa. The way people perceive their choice of water sources is indicated for one household in a humid section of Ganda territory (north of Lake Victoria) to illustrate the method of ordered choice. The family living at the house has within easy walking distance six sources where water might be drawn. Source 1 is a borehole where pure, mineralized water may be had-for the pumping. Source 2 is the iron roof of the mud-and-wattle house. In the opposite direction a path leads to a spring at 4, a small stream entering a swamp at 5, and another seepage at 6. Somewhat nearer than 4 and along a side path is another spring at 3.

An interview with the woman of the household shows that she is aware of the borehole but does not go to it because she considers that the water, judging from taste and the color it gives tea, would be injurious to her children, and she also distrusts the stream at 5. She also realizes that she could take water from the roof if there were a suitable gutter, drain pipe, and barrel, but that would be too expensive. Source 3 rates high on quality, ease of obtaining water, and distance, but requires going across the land of an irritable neighbor and thus she chooses 4, in preference to the furthest source at 6 (White, Bradley, and White, 1972).

process by which people assess risk and choose among alternatives in dealing with it. These are reviewed in more detail by Slovic et al. (1974) and are noted where appropriate here.

Field observations on natural hazards are available from more than forty places in seventeen countries (see map, pp. 264–265). Conducted since 1965 as a part of a collaborative program of the International Geographical Union's Commission on Man and Environment, the field studies tell what kinds of people adjust to different hazards in what kinds of ways. They are only the beginning of an examination of the choice process and raise more questions than they answer, yet they do outline some of the patterns of hazards and adjustments in roughly comparable form. Field observations in ten countries are described separately (White, 1974), as are the air pollution studies in the United Kingdom (Burton et al., 1974). In addition, studies in a few other countries provide analysis that is reported here.

Specifically, comparison is made of findings from local studies of tropical cyclones in the United States, Puerto Rico, and the Virgin Islands; droughts in Australia, Brazil, Mexico, Nigeria, and Tanzania; floods in India, Malawi, Sri Lanka, the United Kingdom, and the United States; volcanoes in the United States; and coastal erosion, urban snow, and high winds in the United States. Comprising thirty sites and interviews with approximately thirty-three hundred households, these studies reveal the diversity and some of the similarities among adjustments that are under way (see Table 4.1).

In addition, use is made of a wider range of studies (described in Table 4.2), many of them precursors of the comparative studies or investigations that were made at the same time but do not lend themselves to the same quantitative examination as do the twenty studies noted above. While they do not fit into the kind of analysis reported for the comparative sites, they provide penetrating examination of local situations and throw light on the individual choice process.

Although five of the comparative studies in Table 4.1 involve habitat where the vulnerability to extreme events is uniform, the others embrace a wide range of risk. All but three suffered a damaging event, some as recently as within the preceding year; severe cyclones had not struck later than 1942 in the Virgin Islands and 1956 in Puerto Rico, and the San Francisco earthquake of 1906 had not been repeated by major tremors thereafter. In most of the sites the worst event of memory had occurred some time after 1956. The frequency of damaging events was high in a few places and rare in others. In contrast to San Francisco and its decades without serious earthquakes, residents of Boulder reported high winds on an average of three times a year, and farmers in the Ganga floodplain noted floods once in five years on the average.

By design, the people interviewed were mostly men who were heads of their households, between the ages of 37 and 54. The family size ranged from 2.3 in San Francisco to more than 15 in northern Nigeria. They varied in education from almost wholly illiterate pastoralists in northern Nigeria to sites in the United States

TABLE 4.1. Comparative Study Sites

Type of hazard	Location	Household number	Land use	Occupation	Age of respondent (average)	Number in household (average)	Tenure (% own)	Illiterate (%)
Flood	Rock River, Ill.	149	Urban recreation	Retired management	47	—	81	3
	Sri Lanka	241	Farm	Farmer	55	6	58	19
	India	66	Farm	Farmer laborer	38	—	85	49
	Malawi	120	Farm	Farmer laborer	48	—	Shifting	55
	Shrewsbury, England	132	Urban	Artisan, retired	52	3	38	0
Hurricane	Galveston, Tx.	120	Urban	Mixed	48	—	71	3
	Pass Christian, Miss.	120	Urban	Mixed	43	4	78	0
	Tallahassee, Fla.	120	Urban	Mixed	44	5	73	0
	Puerto Rico	147	Urban	Mixed	43	5	69	5
	Virgin Islands	93	Urban	Mixed	44	5	61	9
Drought	Australia	181	Ranching	Farmer	46	—	85	1
	Brazil	397	Fields	Farmer artisan	42	6	62	44
	Nigeria	150	Pastoral	Farmer	50	16	Shifting	91
	Tanzania	474	Fields	Farmer	45	7	Shifting	46
	Yucatan	118	Fields	Farmer	45	7	Shifting	33
Earthquake	San Francisco, Calif.	120	Urban	Mixed	46	—	68	0
Snow	Marquette, Mich.	120	Urban	Mixed	38	4	57	0
Wind	Boulder, Colo.	120	Urban	Mixed	394	90	0	0
Volcano	Hawaii	101	Farm	Farmer mixed	51	4	78	11
Coastal erosion	Bolinas, Calif.	120	Urban	Mixed	41	3	62	0

TABLE 4.2. Other Detailed Studies of Hazard Situations

Type of hazard	Location	Number of households	Chief occupations	Reference to report
Avalanche	Huayies Valley, Yungay, Peru	86	Urban	Penaherrera de Aguila (unpublished)
	Austria—near Innsbruck and Heiligenblut	25	Rural	Bauer (unpublished)
Drought	Eastern Kenya	610	Rural	Mbithi and Wisner, 1974
	Oaxaca, Mexico		Rural	Kirkby, 1974
Earthquake	San Francisco, Calif.	120	Urban	Jackson, 1974
	Cornwall, Ont.	118	Urban	
Flood	U.S.—6 towns	109	Urban	Kates, 1962
	U.S.—16 agricultural	0	Farming	Burton, 1962
Tropical cyclone	Coastal Bangladesh	419	Rural	Islam, 1974
Volcano	Costa Rica	170	Rural	Lemieux (unpublished)
Frost	Southern Florida	100	Rural	Ward, 1974
	Wasatch Front, Ut.	75	Rural	Jackson, 1974

where almost all those interviewed were at least high-school graduates. An effort was made to sample a variety of income levels both within each site and from site to site. Different income levels relative to community norms were sampled in Shrewsbury, England, with perhaps U.S. $2,000 per capita yearly, and in Sri Lanka, where the annual income is less than U.S. $200. The total set of sites included sedentary farmers, shifting cultivators, fishermen, city laborers, artisans, small businessmen, manufacturers, teachers, and government workers.

By and large, they were people who in discussion emphasized the advantages of their location, were aware of the hazard under study, and who had suffered damage from it or knew of damage to others. With the exception of Sri Lanka, their experience with what they considered good years was more frequent than it was with bad years. They were aware of a modest number of possible adjustments to reduce damage, and their exposure to government assistance in time of need ranged from complete independence to heavy dependence upon public aid.

In culture, the sites spanned industrialized Western groups, African and Latin American peasants, and Asians. Buddhist, Christian, Hindu, Moslem, and Animistic religious sects were represented, and forty-one different languages were spoken.

In no sense, however, was the set of study areas representative of the world's population at risk. Nor was it intended to be a statistically valid sample. Rather, it was a selection from different cultures and hazard situations, drawn partly by intent and partly by the fortuitous cooperation of investigators. The results permitted initial probing of the tremendous variety of the earth's patterns of hazard in the environment.

To supplement these field observations there are the findings of previous or additional studies such as our investigations of industrial floodplains in the eastern United States (Kates, 1962, 1965), of air pollution in three British cities (Burton et al., 1974), the Japanese studies of landslides and floods, U.S. studies of frost, and the like (Table 4.2).

When the first detailed field studies of hazard response were made in the 1960s, there was a tendency to extrapolate from observations in a few areas to behavior elsewhere. As more investigations accumulated, the earlier conclusions were seen to be less widely applicable. The initial studies, for example, showed that socioeconomic status was not closely associated with awareness of hazard and opportunities. Later findings were sometimes contradictory, thus indicating lack of generality in the earlier conclusions.

Facing these inconsistencies, it is necessary for us to ask how valid and illuminating were the field observations, what few findings seem to be generally applicable, and what combinations of local conditions account for differences in findings from place to place. In drawing upon these findings the focus is on problems of general import; no attempt is made to summarize the bulky literature on the topic.

APPRAISING HAZARD

People have difficulty in defining the future timing and significance of an extreme event in a fashion that is intelligible to others. Part of the problem is the difference among people in time horizon as dictated by the length of time to which they look forward, and the number of cognitive and affective obstacles standing in the way of accurate individual recognition of uncertainty and probability.

Long-term versus Short-term Horizons

A real estate developer standing on the ground floor of a new apartment building on the floodplain of a creek in a Missouri Valley town was asked whether he thought he was taking any risk in locating a structure there. He replied to the contrary and, when pressed, observed further that he knew that the stream had many years earlier reached a stage at that point as high as his shoulders. How then could he say there was no risk? His answer was, "There isn't any risk; I expect to sell this building before the next flood season."

His response was accurate so far as he was concerned, but not necessarily so for the purchaser of the property or for the series of tenants of ground-floor apartments or for the city government which would feel obliged to provide relief for those people when the next large flood arrived. It patently was not correct in the view of the hydrologist who, having mapped the remnant traces of previous floods and calculated the probability of recurrence of other floods, felt a professional responsibility for describing the hazard zone. Yet, the assertion was in fact accurate for the new owner, who happily managed the building for 15 years without suffering a large flood, or for his tenants during that period. Insofar as people operate on short-term horizons, their appraisal of a future event may be drastically different from that of the scientist taking a long view or of a city engineer estimating direct pecuniary costs over a 20-year amortization period for city bonds.

In fact, the financial arrangements and livelihood expectations range so widely that it is misleading to think of a given event as constituting the same hazard for everyone. Even where the economic base and social base are alike, there are other complications in describing the prospect of a potential hazard event.

Cognitive and Affective Obstacles

All people have trouble drawing valid conclusions from a small sample; even professional psychologists with statistical training tend to overgeneralize and to let false intuition get ahead of formal statistical procedures (Tversky and Kahneman, 1971; Kahneman and Tversky, 1973). No wonder that a hurricane-zone shopkeeper who has seen only one severe hurricane in 20 years has difficulty making an estimate of the likely future occurrence of damaging winds. Few geophysical records are available for longer than 60 years, but the estimated recurrence intervals of damaging events may run from 25 to 1,000 years.

As Slovic et al. (1974) have pointed out, some people fall into what is called the "gambler's fallacy," some tend to rely unduly upon the evidence of a limited but concrete sort, and others follow a conservative "anchoring" procedure. The "gambler's fallacy" assumes that if an event has occurred one year, it is less likely to occur the following year. The hypothetical Caribbean fisherman of Chapter 2 was assumed to increase his probability because of recent exposure. Hydrologists know that if the event is random the probability of its recurring is just the same one year as the next. Two floods estimated to have a probability of once in a century took place on the Housatonic River (New England) during the same summer of 1955. Another flood might have come the following year—or not again for two centuries. For extreme geophysical events that are not random—such as earthquakes—the analysis is different, since the severe occurrence results from accumulated strains that take time to build up again once they have been relieved. But for such events as floods and heavy snowfall, people have a tendency to assume that one severe occurrence heralds a period of respite. This was the case for LaFollette, Tennessee, floodplain dwellers (Kates, 1962).

In trying to interpret information about past events, people also seem inclined to use a concrete piece of information—perhaps from a neighbor—as being representative of a larger, more abstract, body of information. The Australian farmer is exposed to a large body of meteorological and climatological data when he approaches the crucial decision of when to seed his crop—and then proceeds to use a reading from his rain gauge as a basis for choice; or he may sow at the same time each year regardless of conditions, or, more likely, scruff the soil to obtain a rough check of soil moisture to back up an innate feeling that "this is the time" (Heathcote, 1974). Sorting out, absorbing, and applying extensive data is a tough job, and one often avoided by employment of a simpler measure at hand.

Faced with such masses of information, people may follow a process of "anchoring" in which they begin with one rough estimate and then adjust it as more information comes in. They "believe they have a much better picture of the truth than they really do" (Slovic et al., 1974), and this generally leads to underestimation of the phenomena. Starting with a view of the possible severity of drought based on previous experience, they adjust their assessment of any new data toward that anchor. Citrus growers in Florida (Ward, 1974) and apricot/cherry/apple growers on the slopes of the Wasatch mountains in Utah (Jackson, 1974) appear to consistently underestimate the probability of severe frost. Jackson observes that the "probability of total destruction (less than 22°F after April 15) is .32, but only 4 percent of the growers indicate that they expect total destruction once every three years."

It is not certain to what degree awareness of events is affected by the basic cognitive obstacles involved in handling probabilistic information, and how much it is influenced by differences in the cultural and economic conditions of the people involved, or by their experience and value orientation. City dwellers or new arrivals in the countryside may be less accurate in their judgment of natural phenomena primarily because they have received less information by personal experience or from others; but lack of data would not account for views of the events that depart radically from the probabilistic framework in which natural scientists analyze the data. Mitchell (1976) found that newcomers to the New Jersey coast built up relatively accurate environmental information in less than 15 years.

What is generally true is that people living in hazardous areas have different views of the hazard than does a scientist studying the same natural phenomena. They are not necessarily "incorrect" in their appraisals of the events; they pay attention to different characteristics and often deal quite differently with probabilities. Indeed, given their particular needs, they may arrive at more accurate or useful appraisals than the "experts."

As was noted among the LaFollette, Tennessee, floodplain dwellers in 1962 (Kates, 1962), people resort to a number of interpretations of natural events. Supernatural explanations are often given where there are no readily comprehensive explanations of other sorts. Just as a person needing groundwater may be disposed to believe in a water watcher in order to make a choice of a well location (Hyman and Vogt, 1974), the individual confronted with complex hydrological or meteorological events must seek other, operable descriptions.

A determinate ordering of the universe is one device: droughts or floods are seen as recurring regularly and periodically. Or they are viewed as becoming more frequent or less frequent. Denial of possible recurrence of the event is another device. Among twenty sites, respondents' views of the prospect for a return of the extreme event were as shown in Table 4.3. The sophisticated view—that it could happen any year—is the prevalent view especially in high-income areas. The deterministic view—consistent with the gambler's fallacy—that the phenomenon is grouped or regular in occurrence is less common but strong in pre-industrial societies and areas of yearly flooding. Taken as a whole, field studies indicate that most residents are informed and knowledgeable, but our site observations are not uniform in that respect.

Boulder, Colorado, with its annual wind storms, shows a high emphasis on regularity, and Bolinas, California, cliff dwellers subscribe to the grouping explanations. For the most part, Nigerian

TABLE 4.3. Views of Future Events in Comparative Sites

	Percentage of respondents to a story about a hazard event				
Site	They don't know	It could happen any year	It happens in groups	It returns regularly	It won't happen again
Floods					
Rock River, Ill.	4.0	76.5	11.4	7.4	7
Sri Lanka	—	78.4	7.9	13.7	—
India	3.0	63.6	15.2	18.2	—
Malawi	8.3	42.5	22.5	21.7	5.0
Shrewsbury, England	4.5	72.0	2.3	4.5	16.7
Tropical cyclone					
Galveston, Tx.	6.0	85.5	1.7	6.0	.9
Pass Christian, Miss.	—	90.8	2.5	6.7	—
Tallahassee, Fla.	—	95.0	1.7	2.5	.8
Puerto Rico	2.7	76.7	3.4	15.1	2.1
Virgin Islands	3.3	84.8	3.3	6.5	2.2
Drought					
Australia	1.8	81.3	6.4	10.5	—
Brazil	5.4	48.8	22.3	6.4	17.1
Nigeria	1.3	16.8	4.0	3.4	74.5
Tanzania	.3	64.0	9.0	13.0	13.8
Yucatan	1.7	45.8	18.6	28.0	5.9
Earthquake					
San Francisco, Calif.	9.2	75.0	10.8	4.2	.8
Snow					
Marquette, Mich.	.8	84.2	4.2	10.0	.8
Wind					
Boulder, Colo.	2.5	67.5	7.5	22.5	—
Volcano					
Hawaii	11.0	77.0	7.0	4.0	1.0
Coastal erosion					
Bolinas, Calif.	6.7	39.2	46.7	6.7	.8

farmers respond to questions about drought by denying that it will return—a stance that is embedded in belief that the occurrence of dry periods is divinely ordered and is neither to be predicted nor protected against. Denial of the event is reported in numerous other locations, although nowhere so fully as in northern Nigeria. At Shrewsbury, England, the position that floods will not come again is associated with higher age and lower income; in recent Texas studies, however, no association was found among age, income, and expectation of flooding (Baumann and Emmer, 1976).

When people are asked to recall severe events of the past, they have been found in a few instances to conform to two patterns. The most recent event tends to be regarded as more severe than equally great events that occurred earlier. And the most recent severe event tends to obscure the accuracy of recollection of preceding events. Studies of drought descriptions in two areas of Mexico—Oaxaca and Yucatan—show that the peasant close to the land may be accurate in dating the last major event, and that details are more hazy for preceding events (Kirkby, 1974; Parra, 1971).

CHARACTERISTIC APPRAISALS AND CHOICES

The variety of human activity—and of natural events that may interrupt it—is so immense that there might be expected to be little behavior in the face of natural hazard that displays easily stated regularity. Still, a few generalizations do seem to apply. At particular places a limited number of factors seem to interact with distinctively local consideration to influence the choices made by individuals. The common elements in these choices leave us with exciting and perplexing questions as to the conditions in which people will respond in future to the rising number of uncertainties that attach to human use of resources resulting from new technologies and capacities to manipulate the environment.

Hazard Appraisal

It is not uncommon to hear people living in a hazardous place— especially if that place be a large urban area—declare that they are free of risk in their location. They may know of the threat of an extreme event and not regard it as a hazard, as, for example, dwellers on the flanks of the Puna, Hawaii, volcano knew of eruptions 10 years earlier and did not count such as a danger (Murton and Shimabukuro, 1974). This is rare in rural areas, where observations show the farmer as keenly aware of the details of hazard risk.

In urban areas, where there is greater mobility of population and less direct contact with the land, the occupant of hazardous terrain may be less informed. The more acute case is found in rural areas newly settled by town folk who are sometimes genuinely unaware of hazards created by their very occupance of the area. Chalets are constructed in the path of avalanches in Austrian mountains by city people who do not recognize the tracks of disaster left by earlier avalanches (Bauer, 1972). Exurbanites moving into northern New Jersey farming areas may unwittingly place their houses in floodplains (Beyer, 1969).

In general, urban dwellers appear to be less sensitive to the possibility of extreme events than rural people confronting the same phenomena. In Sri Lanka, more than 60 percent of the urban residents of selected flood zones are uninformed of the number and height of floods, whereas their country cousins can recall with accuracy the dates and magnitude of overbank flows for the preceding decade (Hewapathirane, 1977).

One consideration that may be related to the urban–rural difference is that particular natural hazards in particular places are perceived as only one of a set of possible events — including social hazards of automobile traffic, crime, and the like — that threaten the security and well-being of the population, as exemplified in the results of tests carried out in Toronto (Golant and Burton, 1969). It is common for residents, like the farmers on the flanks of Puna, to regard other problems as far more important. Thus, homeowners along the eroding shore and unstable cliffs of the California coast near Bolinas consider that coastal erosion is overshadowed by the possibility of earthquakes. Rowntree (1974) finds that they view coastal erosion as a persistently nagging environmental nuisance that is merely symbolic of other hazards. Although they are aware of landslides and cliff failures that could destroy some of their properties, 80 percent "don't know" when asked what kind of damage might result from further erosion of the shore. Among all the sites studied, in eleven sites more than one-half of those interviewed named as important some problem other than the hazard in question.

This is consistent with findings of surveys of attitudes toward air pollution. Residents of an area visibly clouded by particulate matter and subject to heavy loads of other pollutants may place little emphasis on dirty air by comparison with other troubles such as runaway inflation or rats or safety in the streets. The acuity of perception of a hazard in nature is partly a function of the mix of the need for the resource and the social problems confronting people in that place. In the Ganga floodplain of India, where the sole source of income is from that land, the rising water may be a matter of the most intense concern. For owners of residential property along the Rock River in Illinois it is of secondary importance (Moline, 1974).

The more frequent the experience of the individual with the extreme event, the more likely is the estimate of its recurrence to accord with the statistical probabilities. This, however, is subject to conditions imposed by factors of age, economic situation, and personality. Awareness of the risk of extreme events varies with the frequency of the event but also with the magnitude of the most recent events. The regularity or periodicity of occurrence of the extreme event enhances the relation between frequency, magnitude, and awareness. The greater the periodicity the more acute the awareness, as shown in the Ganga and Malawi floodplains.

In urban settings, owners are more sensitive to hazard characteristics than tenants. Thus, the homeowner in the hurricane zone of Galveston is more acutely aware of the prospect for another surge than is the renter. In rural settings the opposite seems to be true. Probably because he has more at stake, the tenant is more likely to be sensitive to subtle differences in severity, frequency, or timing. It is common to find the reported recurrence of an extreme event consonant with the obligations and expectations imposed by livelihood. As Adams (1971) showed clearly for people going to the bathing beach on a summer day, the estimate of the likely severity of a predicted storm decreased as the summer waned and the opportunities to visit the beach diminished.

Acuity of perception of the extreme event probability increases with age up to some point in middle life, but beyond that point no further locational decisions are anticipated, as Saarinen (1966) found with wheat farmers in the U.S. Great Plains. After that period, acuity seems to decline.

Perceived Alternatives

Many people are reluctant to state that there is nothing that they or their government can do to prevent damage from the hazards they face in nature. On the other hand, Nigerian farmers, who (as we have seen) deny the recurrence of a drought, may assert that they are powerless to take other action. A Malawi peasant who expects a serious flood may say that he can do nothing to prevent damage from it. People who declare that they do not know of action that could be taken in response to a tornado warning display distinctive personality traits such as lack of responsibility for directing the future. Similar responses are found more frequently in areas with low severity such as on the Tallahassee, Florida, coast, in places with very intensive but rarely occurring events (such as a California earthquake zone), and low-income peasant cultures where the mode of loss tolerance has not been abandoned (Table 4.4).

In contrast to the Nigerian and Malawi farmers, small landowners and tenants in Sri Lanka and India envision numerous steps that they can take. The more common set of alternatives has to do with emergency action. We find many people who realize that they can evacuate critical areas and take precautions to protect lives and

TABLE 4.4. Modes of Response in Comparative Sites

Type of hazard	Location	Believe hazard causes trouble (%)	Regard effects as not significant (%)	Expect damage to be total or substantial (%)	Believe people can do something to reduce damage (%)	No mention of any positive action (%)	Mention preventive action (%)	Mention move or change use (%)
Earthquake	San Francisco, Calif.	37	30	36	56	50	2	0
Hurricane	Virgin Islands	34	74	44	66	9	1	0
Flood	Malawi	85	6	89	7	6	0	0
Drought	Nigeria	99	0	100	22	18	0	0
Volcano	Hawaii	40	7	80	12	0	2	1
Flood	Shrewsbury, England	45	17	27	44	6	3	—
	India	100	0	94	65	4	17	0
	Sri Lanka	88	0	89	98	—	—	—
	Rock River, Ill.	74	19	24	36	13	44	0
Hurricane	Tallahassee, Fla.	8	6	58	37	5	1	0
	Galveston, Tx.	72	60	28	94	1	4	0
	Pass Christian, Miss.	100	0	72	43	2	2	0
	Puerto Rico	67	16	65	—	2	1	0
Drought	Tanzania	93	—	79	—	19	47	11
	Yucatan	100	1	95	51	16	11	0
Snow	Marquette, Mich.	85	15	7	74	10	22	0
Wind	Boulder, Colo.	100	27	32	85	10	7	1
Coastal erosion	Bolinas, Calif.	94	26	52	74	72	19	2
Drought	Brazil	84	15	—	—	10	38	31
	Australia	88	00	—	—	8	16	10

property. They are aware of the possibility of seeking other income sources and of taking steps when flood or hurricane or drought peak impends. As reported in Chapter 2, the diversity of action is immense. Those affected who have reached the threshold of counteraction are able to specify what could be done. Whether or not they do so is another matter.

A comprehension of the range of remedial or preventive activity depends in part upon confidence that the coming of an extreme event can be foreseen. Although some groups, such as Nigerian, Brazilian, and Australian farmers and San Franciscan fault-zone dwellers, are generally doubtful that the coming of the next event can be predicted, other groups in substantial numbers believe that they can tell the signs of its impending arrival.

One such group is the peasants who rely upon individual or collective village insight. Yucatan maize growers have confidence in their reading of early winter weather as a presage of rainy days during the cultivation season. Sri Lanka paddy workers think they can foretell the date of the next flood, and they are practiced in inferring, from upstream rainfall in their small drainage basins, when their rivers will go out of bank.

In contrast, from 30 to 60 percent of the residents of city areas assert that the government meteorological service will help them anticipate the coming of snow, wind, flood, or hurricane. Among dwellers in the Shrewsbury floodplain, about 40 percent doubt that there is anything they can do to reduce loss, but 62 percent believe that the government will give them some warning.

Perception of alternatives also is involved in the expressed view of where people can go for help as a result of a disaster. A substantial proportion (45–60 percent) of Ganges plain and Yucatan peasants say that there is no one to whom they can turn. Similarly independent are Michigan city folk, who have no precedent for government assistance after a heavy snow. The northern Nigerian looks to Allah for succor. From 5 to 20 percent of those in other areas go to family and friends. In most areas, however, the tendency is to turn to either government or special groups such as the Red Cross and the Church. Low-income areas place heaviest emphasis upon government; high-income areas rely more on voluntary, specialized sources of assistance.

The extent to which an individual crisis situation causes a change in the perception of alternatives is not clear. Often there is a significant difference in the number of alternatives when an individual is forced by rude circumstances, rather than by communicated information, to canvass ways of reducing loss. On the basis of field studies in a few areas, the number of adjustments perceived may be expected to drop off as the use-location threshold is crossed and the individual concentrates on moving out of the hazardous place or on making radical changes in resource use.

For industrial firms it reportedly requires a crisis to institute corrective operations (Cyert and March, 1963); government agencies for their part are often thought to proceed by comparing the present adjustment with one or two analogous adjustments rather than by canvassing the whole theoretical range of possibilities (Lindblom and Braybrooke, 1963). To the degree that they follow such alleged procedures, people may be expected to examine a narrow range at any one time and to move the range along the wider spectrum, like the illuminating beam of a flashlight slowly scanning a scale in the dark.

Adoption of Adjustments

In assigning values to the expected consequences of taking a particular action in a particular state of nature, people are biased in the direction of laying heaviest stress on a single dimension and are hampered by the limitations of their experience. They tend to minimize the expected losses, but the value they place upon loss of time or of property or of income differs greatly according to personality traits and to cultural, economic, and social constraints, including the role played by government. In controlled laboratory conditions people have trouble being consistent in expressing their preferences when obliged to choose, and they are biased toward laying heavier weight on outcomes that can be measured for all the alternatives being evaluated (Slovic et al., 1974).

It is relatively easy to ask people what they think they will do when the event occurs. In a survey conducted by Becker (1974) at State and Madison Streets in Chicago, the busy shoppers estimated that if an atomic bomb landed in the "Loop" it would kill 97 percent of the residents of Chicago. Yet, when it came to predicting what they might be doing three days after the bomb fell, more than 90 percent figured that they would be helping to bury the dead or taking care of themselves, only 2 percent believing that they would have died in the blast. Similarly, 96 percent of the citizens interviewed in San Francisco thought that an earthquake might occur. As for the consequences, about 65 percent stated that they did not expect more than slight damage. This is what people say; it is far more difficult to discover what in fact they believe or might do.

Instances are many in which people use criteria other than economic optimization to judge outcomes. Peasants in the cyclone-torn shore zone of Bangladesh, for example, place heavy weight on maintaining family and community ties, and are willing to sacrifice income advantages and even to risk life in order to remain in congenial situations (Islam, 1974).

People who regularly take action before the extreme event are likely to be those in intensive agriculture who feel that they can predict its coming, or in cities, where they rely upon government forecasting services and warning. In either case they are confident of capacity to foresee the event.

Although the total number of adjustments chosen by groups who decide upon some kind of counteraction is large, the average number adopted by individuals is relatively small (see Chapter 2). The mix of adjustments for an area aggregates the individual choices and embraces considerable diversity.

Only in a few areas do substantial numbers of people contemplate radical changes in location or resource use. Brazilian peasants move away from the drought polygon in large numbers. Australian farmers are willing to consider it as a solution. Cliff hangers in California or volcano dwellers in Hawaii realize that they face permanent evacuation as a possible eventuality, but they rarely move before required by physical destruction of homesite.

Factors in Choice

In choosing among adjustments, at least four classes of factors are found to be associated with the decision in a number of areas.

First, prior experience with the hazard is commonly linked with the choice made. The degree of severity of the risk, and the length of exposure to its consequences, are often related to the type and number of adjustments selected.

Second, the material wealth of the individuals concerned is usually associated with the extent to which they take risks. Wealthier peasants are more likely to experiment with a variety of measures against flood or drought, and they have both the necessary pecuniary support and the security of a carry-over to the following year.

Third, personality traits have been found to figure in certain choices, particularly in the face of severe, intensive events. Baumann and Sims suggest a direct relation between a sense of inner control and response to tornado warnings (1972), as well as a choice of hurricane adjustments (1974). Somewhat similar factors seem to be at work in the choice of radical use-location adjustments in Brazil.

Fourth, the perceived role of the individual in a social group may be influential. Most field observations give support to a condition — sometimes regarded as a dilemma — that prevails in situations of social change. While people seem enmeshed in an inertia that discourages any departure from the existing set of adjustments to the hazard, they still have the capability of very rapid adoption of new adjustments when circumstances are favorable. City officials can be quick to take advantage of technological innovations such as cloud seeding, even though its effectiveness is largely unproven; traditional Kenyan farmers can shift in a few years to drought-resistant crops; a flood-warning system can be enforced with alacrity in a Japanese village.

In explaining such behavior, the perceived roles of individual and government may be highly influential. In Heberlein's theory (1971), a major component of any choice is the sense of responsibility that the individual has toward the causes of the situation and the possible remedial action. According to this view, what the individual does is strongly related to his recognition of his capacity to act and his sense of social responsibility to do so. The sense of individual capacity in turn is related to the sense of efficacy, of knowing what is to be done and when.

In reviewing the variety of factors at work, we have reason to suspect that many firms and individuals respond to the complexity of choice by adopting simple, unambiguous decision rules. A resident of a tornado zone says, "I only go into the cellar if I can see the funnel"; he doesn't try to judge whether the radio warning warrants such action. Large business firms reduce uncertainty in similar fashion by enunciating rules that can be applied without discussion (Cyert and March, 1963).

From the observations so far made, it would appear that individuals have substantial tolerance to stress created by the risk or by the uncertain timing of natural events. In making adaptations over long periods of time they are not willing to pay heavily to eliminate the hazard. In the short run they do not seem to seek risk for risk's sake, as would a man who courts danger by parachuting solely for pleasure. Neither do they place a high negative value on stress. There appears to be a greater propensity to assume risks knowingly among those who are relatively secure economically; the poor tend to choose more cautiously. Older people who later in life become committed to a stressful situation may also be less flexible than the young in canvassing stressful options. Differences in personality traits with regard to seeking stress appear to be relatively unimportant in ordinary circumstances. In extreme situations, however—when the tornado funnel looms black on the horizon, when the drought drags into its third parched year—personality traits may wield more influence, and it is then that the tolerance of stress may influence the decision to retreat to the cellar or to leave the sun-baked farm forever. Thus, the stress-seeking or stress-reducing inclinations of people appear to play an influential role in the tails of the frequency distribution and to figure little in the bulk of ordinary decisions.

FOUR BEHAVIOR PATTERNS

When the behavior and attitudes of groups of people in a hazardous place are examined, they are seen to follow at least four patterns. Pattern one is that found in the zone of a potential earthquake or along the shores of a coast rarely traversed by a tropical cyclone. The majority of the inhabitants either deny the risk or assert that they do not regard the hazard as a problem for themselves or their neighbors. A substantial number dismiss the probable effects as not significant. This describes the situation on the Virgin Islands coast and on the San Andreas fault (see Table 4.5). The principal responses to remote events are adaptive insofar as they take account of the hazard; their fate is determined by the absorptive capacity of their livelihood.

A second pattern of response is that displayed by dwellers in the path of Malawi floods, Nigerian drought, or Hawaiian lava flows. These have crossed the threshold of awareness of the hazard regard its effects as significant and judge that there is little that people can do to reduce its impacts. They tolerate the prospective loss without taking countermeasures. Their response is passive: the chief adjustments adopted are to evacuate or to seek help.

A third pattern is apparent among the greater number of sites studied. In floodplains other than Malawi, in several of the drought areas, and in snow and wind sites, people believe that action can be taken to reduce loss from the hazard. They know that damage can be substantial or total. They cross the threshold of action. Adjustment in large part is emergency action upon receipt of warnings, although in some areas there is serious effort to anticipate and prevent losses. Where the hazard is especially pervasive, as in coastal erosion, many people regard prevention as the only positive action.

The fourth mode is illustrated by Brazilian and Australian drought areas, where a significant proportion of the people expect to

TABLE 4.5. Characteristics of Four Behavior Patterns

Response mode and location	Event characteristics				Societal characteristics			
	Extreme event		Experience		Resource use		Material wealth	
	Intensive	Pervasive	Recent	Remote	Extensive	Intensive	Low	High
Absorb								
San Francisco, Calif.	+			+		+		+
Virgin Islands	+			+		+	+	
Accept								
Malawi	+		+		+		+	
Nigeria		+	+	+		+		+
Hawaii	+			+		+		+
Shrewsbury, England	+		+			+		+
Reduce								
India	+		+			+	+	
Sri Lanka	+		+			+	+	
Rock River, Ill.	+		+			+		+
Tallahasee, Fla.	+		+			+		+
Galveston, Tx.	+			+		+		+
Pass Christian, Miss.	+		+			+		+
Puerto Rico	+			+		+	+	
Tanzania		+	+				+	
Yucatan		+	+		+		+	
Marquette, Mich.	+		+			+		+
Boulder, Colo.	+		+			+		+
Bolinas, Calif.		+	+			+		
Change								
Brazil		+	+		+		+	
Australia		+	+		+			+

move away or change their land use. They regard prospective effects as significant and commonly take preventive action, but they are prepared to consider drastic changes in location or livelihood.

The sites are arranged in rough order according to the prevailing mode of action and attitude. They have certain other characteristics in common as shown in Table 4.5.

Does this help in understanding ways in which Hamisi Juma and Bill Bauer (whom we met earlier in this chapter) severally approach the three thresholds of awareness, action, and use location? Clearly, our two farmers rarely begin any serious canvass of events or possible adjustments until the damage becomes sufficiently severe to mark the natural event as memorable and possibly important. At that stage they are both aware of the hazard and ascribe some explanation to its pattern of occurrence. Once the loss threshold is reached, the appraisal of adjustments begins. At that stage they are each affected by their perception of alternatives and of the likely consequences. If none of the alternative adjustments, short of migration, meets their basic criteria, the use-location threshold is crossed and the choice then is among possible alternative locations or livelihoods.

IMPLICATIONS OF INDIVIDUAL CHOICE

The rising toll of property loss from most of the natural hazards that complicate man's life stems generally from decisions by individuals acting within constraints established by their governments. It is individuals in the aggregate who account in large measure for increased exposure by Australian farmers to drought loss, for new construction in Sri Lanka floodplains, for occupation of low coastal areas in Bangladesh, and for housing to the windward of English factories. While some decisions affecting direct public purposes are made by public agencies—like locating the Kennedy Center for the Performing Arts within the range of the 1 percent probability flood from the Potomac River in Washington—the greater number of the critical choices are the work of private persons influenced profoundly but indirectly by their governments.

In the process of reaching these decisions, a few elements seem to be widely shared. Our concept of the decision process—as one involving awareness of the hazard and of possible courses of action as a base for appraisal and selection of alternatives—provides a framework for describing the greater number of factors entering significantly into this set of transactions with the environment.

Although the ways in which particular factors affect the choice in particular situations are understood imperfectly, there is enough insight into their workings to yield a few generalizations as to the relation of the individual response process to trends in the group response to hazard.

The prevailing patterns of individual behavior in industrial societies offer no assurance that without strenuous effort by public agencies the rising loss from infrequent but catastrophic events will be stemmed. These patterns include the personality characteristics of individuals, their perceptions of hazard, the short-term horizon usually adopted, and the situations in which they are obliged to make choices that enlarge susceptibility to rare events while adjusting with less pain to more frequent events. Such patterns of choice are not the inevitable result of modernized societies. Nonetheless, insofar as group or governmental influences are absent or are poorly designed, they presage increasing suffering from rare events.

It is easy to read into every human interaction involving resource deterioration the conclusion of the "Tragedy of the Commons" (Hardin, 1968) — that individual choice unconstrained by firm government regulation will lead to destruction of the common resource and the society. The record suggests, however, that as the risks of such deterioration mount, a variety of measures will be taken to counter them. People make a wide range of adaptations and coping adjustments that may retard or reverse destruction and enhance productivity. In dealing with extreme events in nature, more intense development at first restricts the wide range of these individual adaptations and adjustments in favor of collective adjustments. To that extent it helps breed catastrophes while moderating the effects of less threatening events. But there also is evidence that critical points may be reached beyond which the trend may be reversed.

In these as in other issues the focus is on individuals. They are viewed as operating within limits set by their communities or nations which, in turn, face collective decisions in responding to hazard and in guiding the actions of their individual members.

Collective Action

Bill Bauer and Hamisi Juma are not islands unto themselves. They and their land are linked to their neighbors, to kin, to communities, to institutions, to markets near and far. These linkages complicate their lives and livelihood, make them less secure as they depend on the uncertain actions of others, make them more secure as they spread their risk and draw sustenance from the mutual aid and efforts of others.

In his everyday existence, the life of Hamisi Juma is more a collective one than is that of Bill Bauer. His immediate family is larger; his extended African family has no equivalent in Bill Bauer's Australian culture. Both participate about equally in collective religious and political life, market their crops through organized cooperative or corporate institutions, and look to community, region, and nation for certain services to which they also contribute their support. The array of such services Bill Bauer draws upon is larger than that of Hamisi Juma, and he is less self-sufficient than Juma. His society is more specialized and diversified and his links to the collective society are more distant and complex.

In dealing with drought, the wealth and large-scale technical and collective experience of Australia offers to Bill Bauer some adjustments unavailable to Hamisi Juma: massive irrigation works, continuous breeding and distribution of new seed varieties, assistance in moving or providing fodder for livestock, and insurance. These adjustments share the general qualities of requiring large capital investment and extensive preplanning — resources still in short supply in Tanzania. But for both, in the long run, more and more hazard adjustments involve increasing efforts at collective action.

COLLECTIVE LIFE

Guides, Managers, and Services

Because adjustments such as flight before a storm surge are taken by individuals on a Caribbean shore, they are not inappropriate for collective action or governmental control in the Bay of Bengal. The choice of action taken by individuals and by collectivities, including governments, is strongly affected by their mutual sense of responsibility and their expectations of each other. An individual householder may be disinclined to take precautions against a snow avalanche because he believes that they should be carried out by the community or that community views or policies are not favorably disposed to such adjustments. Communities (Boulder, Colorado, for one) in turn may be discouraged from adopting a new approach to damage from high winds because they anticipate that the appropriate response will not be forthcoming from citizens. Contrariwise, policies they do adopt may fail for want of citizen cooperation or because of unsuitable response.

In pursuing a policy of hazard reduction, community authorities have three broad areas in which they can function. First, by setting laws, regulations, incentives, and penalties they provide guides to the choices made by citizens in a "private" capacity. Second, they make decisions concerning the use of resources and response to hazards, including the deployment of large-scale, expensive technology, such as dams, that can be undertaken only in large, indivisible applications. And third, they dispense hazard adjustment services.

Social Guides

Public decisions guide managerial or private-sector choice in both purposeful and incidental ways. The purposeful guides in the main are intended to protect individuals, reduce damage potential, and to discourage managerial acts that deliver injury to others. The influence on managerial decision can range in effect from the actual co-opting of the individual choice by some public action, as for example mandating them by law or regulation; encouraging them by subsidy, bribe, or exhortation; discouraging them by tax, sanction, or social disapproval; or facilitating them by providing informative advice and the necessary requirements for choice.

Some examples of how these guides operate for three hazards are shown in Table 5.1. To cope with drought, farmers in Tanzania are frequently ordered by regional commissioners to plant a minimum amount of cassava as a famine food, are encouraged to plant early for optimal moisture in cotton-growing regions, and are facilitated in their selection of irrigation by the recently created Smallholders Irrigation Advisory Service. For occupants of U.S. floodplains the government may pre-empt the choice by expropri-

TABLE 5.1. Examples of Governmental Guides to Managerial Choice

Guides	Tanzania drought	U.S. floods	U.K. pollution
Purposeful			
Pre-empt	—	Transfer of flood plain land to public ownership by eminent-domain proceedings	Planning permission refused for new housing estate near heavy metal smelter at Avonmouth
Mandate	Regional regulation for minimal planting of famine-relief crops	Local regulation of building elevation	Factory emissions regulated by the Alkali inspectorate
Influence	Exhortations for early planting by political leaders	Subsidized flood insurance premiums	Subsidy for domestic heat conversion to smokeless fuel
Facilitate	Provided advice by Smallholders Irrigation Advisory Service	California State subdivision deed requirement informing of hazard	Little or no information provided to facilitate choice of individual adjustments
Incidental	Expansion of education for children may encourage shift to drought-prone but bird-resistant crops	Income tax deductions for flood loss; discourges movement out of flood plains	Miner's allowance of free coal; discourages shift to smokeless fuels

ation of the land (with compensation) and converting it to different use. Alternatively, a government agency may require the elevation of buildings above a certain level, encourage people to stay in floodplains by subsidized insurance premiums, and provide information as to the hazard by amendments to deeds in new housing subdivisions, as in California.

In the United Kingdom, factory emission of air pollutants is regulated by the Alkali Inspectorate, and planning permission can be used to guide development into suitable locations. For example, planning permission was refused for a new housing estate to have been built not far from a lead, zinc, and cadmium smelter that has been causing air pollution problems at Avonmouth (Whyte, 1977). Conversion of domestic heating away from the coal fireplace to units capable of burning smokeless fuel are encouraged by subsidy from local and national governments. Little or nothing is done to provide the general public with information on levels of air pollution, as is the practice in several Ontario cities (Burton and Auliciems, 1972), to enable chronic sufferers of bronchitis or other respiratory diseases to choose to stay out of certain areas during periods when pollution levels are high. Such information is often intentionally withheld, and policy does not facilitate the choice of other possible individual adjustments. Related policies have an incidental effect: for example, the National Coal Board practice of allowing miners a quantity of free or so-called "concessionary" coal is one factor in a complicated situation discouraging the shift to smokeless fuels in coal mining areas.

In other countries, actions of public bodies also frequently affect the choices others make in unforeseen and incidental ways. One explanation, albeit a minor one, given for Tanzanian farmers' shift from drought-resistant millet to drought-prone maize is the relative bird-resistance of the maize husk. Increased education reflected in the hours children spend in school has reduced the traditional labor supply of bird guards. Flood damage losses are part of the general casualty loss provision of U.S. income tax deductions. Such tax subsidies may encourage continual occupance of hazardous areas.

Enterprise Managers

In all parts of the world, public, quasipublic, corporate, or commercial bodies of all sorts use areas of hazard potential as sites of residence or work and as opportunities for resource development and use. The Tanzanian Ministry of Agriculture manages a state farm for irrigated rice production; the Boulder, Colorado, City Hall occupies a floodplain site; the plants of the Central Electricity Generating Board in London, Sheffield, and their equivalents in many other cities in England add to and are affected by the pollution of the atmosphere.

Such collective or corporate enterprises differ little in their resource-using and hazard-related activities from their counterparts under individual management. The state rice farm can waste as much or as little water as its plantation counterpart in other parts of the world; the location decision of city officials in Boulder did not differ from that made by three owners of an upstream motel; and emission rates for publicly owned and operated electric power stations can be no more or less than those from their privately owned counterparts.

Neither economizing for profit nor operating an enterprise as a public responsibility seems necessarily to lead of itself to better hazard adjustment. Industrial firms in Illinois and Massachusetts failed to make what turned out to be highly profitable alterations in plant water supply to reduce water waste until the duress of shortage (Wong, 1968; Russell, Arey, and Kates, 1970). Unaccounted-for losses of upward of 15 percent in public municipal water systems are tolerated for years, despite improved and profitable techniques for locating and repairing system leaks (Howe, 1971a).

Dispensers of Hazard-adjustment Services

A different sort of function, not found in household or individually managed enterprises, is that of providing services that involve hazard response and adjustment. The public or quasi-public agency usually has responsibility for providing one or more services, some or all of which have utility as hazard adjustments. These are the activities that exceed the capability of individual or small groups to organize, buy, or market. These services include those that seek to control or modify natural events: cloud seeding, hurricane barriers, flood-storage reservoirs, irrigation works; those that seek to reduce or prevent damage potential: forecast and warning services, disaster organization, hazard and adjustment research; and those that distribute the damage: primarily, relief activities and insurance schemes. Responsibility for them seems to gravitate toward public organization, whether they are seen as a public need in a planned economy or a nonmarketable service in a mixed-enterprise economy.

Another governmental function related to hazard is slowly emerging in some industrial countries and seems likely to become more widespread: the extension of social security and equity considerations into hazards policy. It emerged in the first instance when questions were raised about the fairness of the distribution of disaster-relief funds. It is applied also to the effects of government acts in different regions of a country, in different sorts of hazards, and in hazard events of the same sort but of differing magnitude.

Inequities in the impact of extreme events are widespread. Those suffering loss from a hazard event in one region may receive government help, while those in another region do not. Farmers whose grain crops are ruined by floods may be compensated, while those whose crops are flattened by wind may not be. People suffering loss in a small-scale disaster affecting only a few people receive less sympathy and assistance than those involved in a larger event. Increasingly, governments are required to demonstrate that certain standards of justice and equity are preserved in relation to loss from natural hazards. Eventually this may lead to governments' accepting the proposition that all citizens should be afforded a minimum degree of protection and financial security from loss from natural hazards however it be inflicted. No government has adopted such a policy yet, but New Zealand has moved in that direction (O'Riordan, 1974).

ADJUSTMENTS

As compared to the decision of an individual resident or user of a hazard area, collective adjustments appear more complex. They may embrace the mixed function of the collective as manager, guide, or dispenser of services, the variations in action and attitude among all members acting within the collective, and the enlarged scale of opportunities afforded by the overlap of individual capability and collective possibility. They share, in common with all adjustments, the decision to bear or share loss, to modify extreme events or prevent their effects, or to change resource use or location. This is illustrated in a wide range of collective forms and experience.

Bearing Loss: The Kenya Famine of 1961

In 1961 a major drought spread across Kenya and parts of Tanzania, causing widespread suffering. In the face of hunger, loss bearing is most real and imminent. For the smallholder farm societies of eastern Africa, there are many factors that mitigate the burden of loss: frequent experience in coping with drought, sustaining religious traditions, and the sense of shared fortune and misfortune that characterize a relatively poor-but-equal society. Yet within such a tradition, great variation in human conditions occurs. When one feels alone in a society where loneliness is rare but where impersonal institutional responsibility is still weak, then the suffering may be unbearable.

James Ngugi (1975), a Kenyan novelist, captures this special poignancy of the collective. He begins with the description of the drought:

> From ridge up to ridge the neat little shambas stood bare. The once short and beautiful hedges — the product of land consolidation and the

pride of Kikuyu farmers in our district — were dry and powdered with dust. Even the old mugumo-tree that stood just below our village, and which was never dry, lost its leaves and its greenness — the living greenness that had always scorned short-lived droughts. Many people had forecast doom. Weather-prophets and witch-doctors — for some still remain in our village though with diminished power — were consulted by a few people and all forecast doom.

Radio boomed. And "the weather forecast for the next twenty-four hours," formerly an item of news of interest only to would-be travelers, became news of first importance to everyone. Yes. Perhaps those people at K.B.S. and the Met. Department were watching, using their magic instruments for telling weather. But men and women in our village watched the clouds with their eyes and waited. Every day I saw my father's four wives and other women in the village go to the shamba. They just sat and talked, but actually they were waiting for the hour, the great hour when God would bring rain. Little children who used to play in the streets, the dusty streets of our new village, had stopped and all waited, watching, hoping.

Among the watchers, Ngugi describes one woman watching not for rain but the agonies of her only son:

. . . As I have said, we had all, for months on end, sat and watched, waiting for the rain. The night before the day when the first few drops of rain fell was marked with an unusual solitude and weariness infecting everybody. There was no noise in the streets. The woman, watching by the side of her only son, heard nothing. She just sat on a three-legged Kikuyu stool and watched the dark face of the boy as he wriggled in agony on the narrow bed near the fireplace.

"Mother, give me something to eat." Of course, he did not know, could not know, that the woman had nothing, had finished her last ounce of flour. She had already decided not to trouble her neighbours again for they had sustained her for more than two months. Perhaps they had also drained their resources. Yet the boy kept on looking reproachingly at her as if he would accuse her of being without mercy.

What could a woman without her man do? She had lost him during the Emergency, killed not by Mau Mau or the Government forces, but poisoned at a beer-drinking party. At least that is what people said, just because it had been such a sudden death. He was not now there to help her watch over the boy. To her this night in 1961 was so different from such another night in the 40s when two of her sons died one after the other because of drought and hunger. That was during the "Famine of Cassava" as it was called because people ate flour made from cassava. Then her man had been with her to bear one part of her grief. Now alone. It seemed so unfair to her.

She leaves the hut to go to the headman of the village, who tells her for the first time that the District Officer (D.O.), a government official,

> rationed out food — part of the Famine Relief Scheme in the drought-stricken areas. Why had she not heard of this earlier? That night she slept, but not too well for the invalid kept on asking, "Shall I be well?"
>
> The queue at the D.O.'s place was long. She took her ration and began trudging home with a heavy heart. She did not enter but sat outside, strength ebbing from her knees. And women and men with strange faces streamed from her hut without speaking to her. But there was no need. She knew that her son was gone and would never return.
>
> The old woman never once looked at me as she told me all this. Now she looked up and continued, "I am an old woman now. The sun has set on my only child; the drought has taken him. It is the will of God." She looked down again and poked the dying fire.
>
> I rose up to go. She had told me the story brokenly yet in words that certainly belonged to no mad woman. And that night (it was Sunday or Saturday) I went home wondering why some people were born to suffer and endure so much misery.

Loss Sharing: The Coming San Francisco Earthquake

The Kikuyu village of Ngugi's tale was not totally isolated. Forty percent of the rations distributed by the District Officer came from Iowa corn farms; the narrator's family received help from brothers working in Nairobi and Limuru. But the linkages were few compared to the interdependencies of highly industrialized societies. The specialization, diversification, enlarged scale, and hierarchical integration that characterize the economy and polity of such societies are described exhaustively for everyday life, but only to a limited extent for extreme events.

One such description has been developed to assist in disaster planning and prevention for the next San Francisco earthquake, which may happen well before the end of this century, earthquakes having been experienced in 1865, 1868, and 1906. A combination of scientific simulation (Algermissen et al., 1972) and reasoned spec-ulation (Cochrane et al., 1974) provides a complex picture of potential loss bearing and sharing.

For the scenario the extreme event is a recurrence of a Richter 8.3-magnitude earthquake on the San Andreas fault, with the approximate isoseismal distribution of intensity of the 1906 San Francisco earthquake. With such an event, there would be maximum intensities IX and X in some areas of San Francisco. Depending upon the time of day such an event would occur, deaths might range from two thousand to ten thousand in the entire Bay Area, the majority occurring in San Francisco proper. There may be as many as forty thousand injured—again depending upon the time of day—and twenty thousand additional uninjured but homeless, depending upon the season of the year and the possibility of fire. Direct property damage might be as much as $6–7 billion (as evaluated by Cochrane, 1974), and an equal amount might be among the indirect economic effects related chiefly to the cessation of economic activity. Of this $13 billion of both direct and indirect damage, approximately 40 percent could be expected to be suffered in San Francisco proper. Who will pay for what might well be the greatest disaster in the history of the United States?

The costs of disaster—the pain of the dying and injured, the disruption of the lives of the living, the loss of real and symbolic wealth—are not easily assessed, counted, or scaled. They are clearly inequitable, as falling heaviest on a few; but the aggregate population affected (whether slightly or severely) is large in an interdependent industrial society. The order of magnitude in burden and population effect is conveyed in Figure 5.1. The costs (the ordinate of the graph) can be read in terms of dollars or as units of social cost; the population affected in some way is the entire U.S. population. The graph illustrates the dimensions of the social burden that would ensue. (While not presented here, a graph can also be drawn for those that gain.)

The seriously injured, the dead, and their survivors bear the greatest cost. The economic value attributed to a life in Western society runs into hundreds of thousands of dollars; the cost of hospitalization and loss of earnings, into tens of thousands.

Much farther down this hypothetical social scale of suffering lie the victims of serious social dislocation that is not or cannot be compensated. We know little about these types of impact, but from analogous situations the costs may be high. The high cost of social dislocation includes the irreplaceable loss of employment, residence, community, treasured possessions, and familiar surroundings. It includes also the loss of replaceable property that is not replaced, or compensable earnings that go uncompensated because of the pride or ignorance of the victim or the inequity and inefficiency of the welfare system of disaster sharing. A review of urban relocation and renewal, of factory closings and declining regions, and of other disasters shows that even in situations where there is great generosity in relief and compensation, the elderly are forced to move,

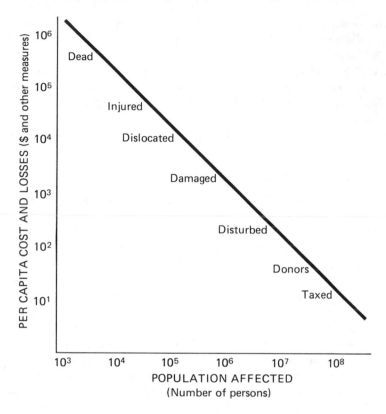

FIGURE 5.1. Loss sharing: Future San Francisco earthquake. Recurrence of an earthquake similar to that of 1906 in San Francisco would lead to a disaster affecting the entire population of the United States. Deaths and injured would be in the thousands, homeless in the tens of thousands, and damages in the billions of dollars. Relief and reconstruction would extend the cost to all taxpayers.

older workers search futilely for another job, the ethnic community is dispersed, and the fiercely independent are left in want. All these suffer a special loss that has never been measured adequately either in monetary or mental-health terms but is nonetheless real and poignant.

In the San Francisco situation, such special victims might comprise the low-income residents of the Mission District, the low- to middle-income residents of the western addition and the low-lying areas of North Beach, and the singles living in multi-story apartment dwellings constructed before 1933.

Farther down the scale are the people who have loss of earnings or of personal or real property, part of which may be compensated for by disaster relief or insurance. With per capita loss from the earthquake in the thousands, their average uncompensated loss may well be in the hundreds of dollars. The list of directly affected includes the surrounding populace whose lives are disrupted in varying degree by the disaster, including some loss of employment. To the extent that the reconstruction process disrupts their lives, they suffer smaller but significant costs. Their number may run into the millions in the densely settled metropolitan Bay Area. More remote are the donors numbering in the tens of millions, who in addition to public assessment share with the victims through private charity and Red Cross contributions. Finally, the entire population of the United States would carry by means of their taxes the several billion dollars of federal aid that would flow to San Francisco for relief and reconstruction.

At the lower end of the scale, the exact pattern of cost-spread is much in doubt. Every great earthquake of this century has had a different pattern of source money for reconstruction. In 1906 San Francisco received no substantial federal aid, but rebuilt with its own funds, those of its fire insurers, and of a small but significant private relief organization. Alaska in 1964 received more in federal aid than the actual damage suffered, but a minimum from insurance and private relief funds. San Fernando, a much smaller earthquake occurring in 1971, received heavy federal aid, little insurance, and little private relief, with a small but significant portion of private loss going uncompensated. The future mix of insurance and disaster aid became a matter of intense investigation by the Congress in 1973. The policy choice was between spreading losses over the insured life of the potential victim or spreading them over the social space of the tax-paying public. Questions as to whether new buildings at somewhat higher initial cost would reduce future costs, and whether older hazardous buildings should be phased out, also began to be debated in federal and state legislatures. Decisions on these matters will shift the burden of future loss patterns.

Modifying Extreme Events: The Liu Ling People's Commune

Characteristic of collective action is to join together in reducing losses by undertaking actions that require capital and planning times beyond the capacity of most individuals. Major progress in reducing the insecurity from natural hazards has taken place in the People's Republic of China. Serious famine, the usual outcome of the long history of floods and droughts, has thus far been successfully averted, but earthquake remains an ominous threat.

Underlying major efforts to cope with drought, flood, and soil erosion are the collective actions of some seventy thousand communes in which over 100 million families live and work. Jan Myrdal (1963), in what is now a classic report, gives life to these achievements in the words of the members of the Liu Ling People's Commune, who live in among the loessial hills west of the River Nan due south of Yenan in northern Shensi. It is difficult country: with adequate water, the loessial soil is capable of high yield, but the rainfall is marginal and the cold season long; drought, frost, hail, and erosion are continuous hazards. As recounted by the villagers, the prospect of collective action on an increasingly larger scale encouraged them to organize into the people's commune. Thus, the old secretary:

> It was in August 1958 that we began preparing the formation of a people's commune. This was discussed at various levels. We had the decision of the Central Committee to go by, of course. The people's commune was set up on the basis of the Great Leap and the need to increase capital investment. Before the people's commune was set up, we had begun building a dam that was to serve three different improved agricultural cooperatives. But it had been difficult getting them to collaborate. The people's commune made it possible to systematize joint effort. We needed a larger unit in order to be able to carry out the big works required. (p. 359)

The young worker:

> We have striven to overcome nature's catastrophes and achieve record crops. We have had great results in our efforts to expand the collective economy, and we have carried out comprehensive water-regulation works. The people's commune is now an entrepreneur, carries out afforestation and herds cattle. (p. 363)
> We have also carried out big water-regulation works. At the end of 1957, we had only 30 mu [1 mu = one-sixth of an acre] of irrigated land. Since then we have built 5 dams, 3 canals and 400 mu of terraced fields, and in 1962 the total irrigated land amounted to 8,320 mu. We have been continuing with this work all the time. I do not have the figures for how much has been done each year. (p. 364)

As the scale of work increases, the commune joins with others in collective action, labor being the key but not the sole medium of exchange, as the young worker explains:

> Larger undertakings, such as water-regulation works, etc., have to be carried out by several labour brigades collaborating. The work is then

managed by a work committee which has its office at the site. The chairman of this committee is always the chairman of the management committee of the commune, while the deputy chairman, who is the one who must always be present at the site, as well as the different labour leaders, are chosen by the labour brigades. Before work begins, the size of each participating brigade's share in the result of the work is fixed. That is, how much of the advantage from the work is for their benefit. In 1958, three labour brigades, Liu Ling Labour Brigade, Kaupo Labour Brigade and Chungchuan Labour Brigade, jointly built the Kaochinko dam. Liu Ling invested 1,900 days work; Kaupo 1,100; and Chungchuan 890.

As Liu Ling Labour Brigade benefited most by the dam, it transpired at the settling up that Liu Ling Labour Brigade owed the other two labour brigades a total of 494 days' work: 394 to Kaupo Labour Brigade and 100 to Chungchuan Labour Brigade. Liu Ling Labour Brigade repaid this by helping in terracing the fields of the other two labour brigades.

In the winter of 1959–1960, Liu Ling Labour Brigade carried out a very big water-regulation project. This cost a total of 14,772 days' work, of which Liu Ling Labour Brigade itself contributed only 4,172. Liu Ling Brigade first repaid a total of 1,568 days' work to the other two brigades which had collaborated, Chungchuan Labour Brigade and the Sun Rises Labour Brigade. They got roughly 800 days' work each. After that, however, it became difficult to repay in days'work, and, after lengthy meetings and much discussion, we agreed that Liu Ling Labour Brigade could pay in cash. Negotiations took a long time. In the end the parties agreed that payment should be at 0.80 Y per day's work performed. That was roughly 0.30 Y below the current valuation of a day's work. It was a matter of a total of 8,352 days' work. Since then no labour exchange project of this magnitude has been carried out between brigades. (p. 373)

Preventing Effects: The Bethlehem Steel Corporation

If the commune represents a basic form of collective action in an agrarian socialist society, the corporation is the most common form of collective in free enterprise, industrialized society. Here again, the need for capital and for preplanning favors larger, diversified units, as reflected in the experience of Bethlehem Steel Corporation, owner of a large, integrated steel mill located on the floodplain of the Lehigh River in eastern Pennsylvania.

In 1942 and again in 1955, the plant experienced major floods with loss in the millions of dollars and production loss of up to three weeks. The 1942 experience served to alert the firm's engineering and production staffs to the possibility of preventing such effects, and they developed an impressive flood-loss reduction system involving warning, structural measures, and emergency measures of all types. In 1965 the system was described by the plant manager as follows (Kates, 1965):

> Having established a method for predicting when and to what extent the river is expected to rise, the next step in formulating this program was the listing of all pieces of equipment in relation to the elevation at which it will be effected by this rise.
>
> Each one of these locations was reviewed and investigated to determine what could be done, if anything, to lessen their vulnerability to the river rises. The topography of the plant is such that water will overflow the banks of the river at an elevation of eighteen feet above normal. After this happens there is very little to be done but wait until it recedes. However, we do have many spots that are vulnerable at elevations lower than this. It was these locations that received our attention.
>
> In many cases, simple barricading or closing-up foundation wall openings was all that was needed. In others it was necessary to install shut-off valves in sewer lines. in still others provision for pumping seepage or gasketing and bolting sewer man hole covers did the trick. All in all we were able to achieve complete protection in some cases, while in others we materially improved elevation levels at which they would be effected. All these improvements afforded us valuable time toward shutting down operations and protecting equipment during our major floods, and provided greater protection against the many other river rises that crested lower than eighteen feet.
>
> Today, new equipment going into the plant is always considered in light of its protection against flood waters, either through water proofing or positioning above the elevation of our major flood crests.
>
> Having established the basis for predictions and area vulnerability, our next step was to organize plant personnel and resources [see Figure 5.2]. Meetings were held with each department head, where we discussed propositions pertinent to his operations. Here a procedure was set up to combat high water periods, the proper personnel and tools listed in addition to when and how this personnel was to be notified of the emergency. (p. 26)

Reflecting on the efficiency of the adjustments, the plant manager concluded:

> In the '42 flood we were caught unprepared and suffered extensive damage. All decisions had to be made on the spur of the moment. We all know, it's one thing to shut down a steel plant in a planned and orderly manner, but it's quite another thing when it catches you unprepared, both from an equipment and a personnel standpoint.

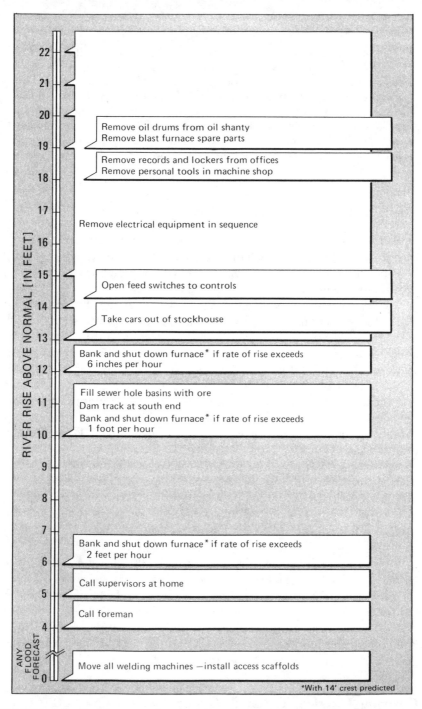

FIGURE 5.2. Contingency plan for blast furnace shutdown, Bethlehem steel plant.

That our flood program has merits was demonstrated in the 1955 flood. Not only were we able to shut down in a more orderly fashion, but we were able to get back into business a lot sooner in many of our operations. This can be attributed to the planning as set forth in the manual and the time interval provided to marshal the required personnel and equipment toward the protection of the more vital areas. This is even more impressive considering the greater amount of mechanical and electrical equipment now in service as compared to 1942. (p. 26)

Changing Land Use: Greater Tokyo

For collectives as well as individuals, fundamental changes in resource use, livelihood systems, or locations are relatively rare, and are undertaken only under great threat of loss. Such threat has been recognized recently in the greater Tokyo area.

Tokyo was severely affected by the Great Kanto Earthquake of 1923. Fires breaking out after the quake killed more than 59,000 people and destroyed some 370,000 buildings. Fifty years later, Tokyo was no less vulnerable than in 1923. It is an overgrown, overcrowded city, 80 percent residential, with many wooden buildings full of facilities where combustibles are stored. Casualties for a recurrence of the earthquake (expected by some Japanese scientists between 1978 and 1994) are estimated as high as a half-million if it should occur on a winter evening with high winds ("Tokyo Earthquake Casualties," 1973).

To reduce the potential toll, major redevelopment of an entire area of the city with drastic changes in land use is envisioned. In the Koto area, six antidisaster open spaces of 50–100 hectares will be developed to serve as refuge from fire. Those open spaces will be used as public parks and athletic grounds and will be surrounded by fireproof high-rise buildings designed to shield the area from fire, as in Figure 5.3. Ringed by a canal network, the disaster open space will appear like a huge, resurrected castle with high-rise buildings forming the walls, canals serving as a moat, and the inner court as ballfields and parks. Building these open spaces will require more than 10 years of effort and about one-third of the total annual Tokyo metropolitan budget—an allocation of funds deemed desirable by the city officials.

Changing Location: Tristan da Cunha

Few fundamental shifts in location can match the odyssey of the people of Tristan da Cunha (Figure 5.4), as they fled a volcanic

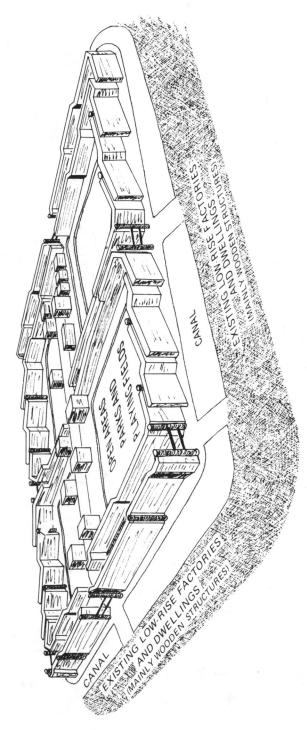

FIGURE 5.3. Disaster refuge planned for Tokyo Metropolitan area. In anticipation of a major earthquake sometime after 1978, redevelopment is planned for the Koto area of Tokyo, an especially vulnerable area below sea level (due to land subsidence) and crowded with flammable factories and residences. The planned redevelopment seems to draw inspiration from the classic walled cities or castles. Open spaces used normally for public parks or athletic fields are surrounded by high-rise fire-resistant buildings, separated by canals from the older existing areas. Such open spaces are intended as havens of refuge for evacuation in the face of fire and inundation from the sea.

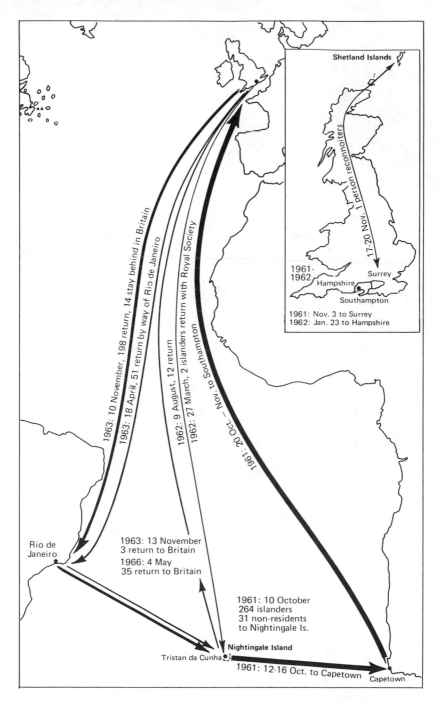

FIGURE 5.4. The odyssey of Tristan Da Cunha. Fleeing a volcanic eruption on their small island home, the islanders of Tristan da Chuna evacuated to England, only to return two years later, unable to adjust to their new location.

eruption on their small (104 square kilometers) island home in the South Atlantic midway between Capetown, South Africa, and Buenos Aires, Argentina.

The 264 islanders, descendants of English, Scottish, American, Dutch, and Italian seamen and African women, cultivated potatoes, with cattle, sheep, and chickens raised by each family for its own consumption; there was some collecting of seabirds, eggs, and guano. The principal economic activities were a rock lobster packing plant and the sale of postage stamps.

After two months of earth tremors in 1961, a volcanic cone broke through the earth's surface in October and led to evacuation of the entire population to an uninhabited island 36 kilometers to the south-southwest. No dissent was recorded. A few days later they were en route to Capetown. The islanders left Capetown for Britain on 20 October and landed at Southampton, where they were taken to a military camp in Surrey.

While some islanders fell sick, others obtained unskilled jobs nearby, and still others went to scout a possible settlement site in the Shetland Islands, the way of life there resembling somewhat their own. In January they all moved to a former RAF base near Southampton. At the same time an inspection party found the settlement on Tristan largely intact and volcanic activity quiescent. England proved uncomfortably hazardous: they were beset by motor traffic, hooligans, aggressive salesmen, and unaccustomed diseases; respiratory sickness and several deaths plagued the group. By the following June, the islanders, after some negotiations, sent another survey party of a dozen men back to Tristan da Cunha. When they reported that settlement was possible, after further negotiations all but fourteen returned in 1963.

The odyssey was not over for all. In 1966, thirty-five returned to Britain, having missed, in the words of an official spokesman, "piped water, electric light, telephones and cars." They were scattered throughout the Southampton area, but the net movement thereafter was back to Tristan. For the remaining returners, the long journey meant, in the words of one, that "if another volcano ever happened to Tristan, they wouldn't get the people off, not so easy as what they did before" (Blair, 1964).

PROCESSES OF CHOICE

When at one critical point in their odyssey the people of Tristan da Cunha voted 148 to 5 to return to their island, it was a rare occasion of direct voting on the choice of location. Seldom are choices as drastic as this one: to stay in a foreign land, or return to a naturally hazardous environment. Because the more routine choices of adjustment are less immediate and pressing, the process by which they are made appears more complex.

THE APPRAISAL OF HAZARD

There is a genuine divergence between individual and group hazard exposure. After the great earthquake of 1963 the planners of Skopje, Yugoslavia, had to decide whether to rebuild the city after it had been virtually destroyed twice previously at roughly 500-year intervals. For them the problem was to weigh the risk in terms of the limited Macedonian alternatives for site (knowing as they did that destruction may come again), and then to rebuild in such a way as to avoid, if possible, the devastation of the past. The sobriety of their decision to rebuild was heightened by knowledge that earthquake would surely, if rarely, come again. But for the individual resident of Skopje, the survivor of 1963 had (on the basis of the historical record) virtual assurance that, except for aftershocks, he and perhaps his children might live out their lives earthquake-free.

This holds true also of less rare and more temporally random events. Given the increasing mobility of societies, the time horizon for an individual moving from farm to city or from one part of a city to another may be very short — for example, less than 10 years. In that case the homeowner might regard a flood with a recurrence interval of once a century as having only a very small chance of occurring during his limited tenure, particularly if he had already suffered one such event during the period (the gambler's fallacy). For the community, however, the recurrence of such an event must be inevitable.

Even without the divergence between group and individual probabilities of hazard recurrence, their perceived experiences may differ considerably. Organizations may have longer memories than their members. Formal procedures for recording experience provide a way of extending the range of individual experience — a significant factor in encouraging coping behavior. Collectives, by virtue of their role, responsibility, or empathy, can extend the group memory to "experience" events that do not directly threaten its members.

This potential for organizational memory is not transferred simply into adoption of adjustments. Few disaster events occur as a total surprise; most are recurrences of previous events — events that may well have been recorded somewhere in the scientific literature, to be "rediscovered" after the latest event. But the potential for drawing systematically on extended experience is there: thus, the cyclone of 1883 serves as warning of still greater potential disaster in Bengal, though no living individual has experienced it.

PERCEPTION AND CREATION OF ADJUSTMENTS

The range of possible adjustments open to collectives is, at least in theory, always substantially greater than that available to individuals and small groups. How are adjustments not only perceived but also invented or created, and by what processes do they become available? Under what conditions are new alternatives added to the repertoire of available adjustment?

Three adjustments that have been adopted in recent years and one that is still undergoing testing are described in order to illustrate the range of possibilities. They are (1) the smokeless area concept incorporated in the U.K. Clean Air Act of 1956; (2) Katumani maize, a drought-resistant crop variety, bulked and distributed first in Kenya in 1964; (3) comprehensive floodplain management made part of the U.S. administrative process in 1966; and (4) tropical cyclone modification by cloud seeding, still undergoing testing and experimentation.

The Smokeless Zone

The notion that smoke from both industrial and domestic sources should be prohibited in designated areas of cities originated with Charles Gandy, a lawyer in Manchester, England, in 1935. As chairman of a voluntary group called the National Smoke Abatement Society, Gandy was concerned to find new ways of curbing smoke levels in urban areas. He developed the concept of a smokeless zone by extension from discussions then being held as to the desirability of noiseless zones—for example, in the vicinity of hospitals, where motorcar horns would be banned. The zone idea was not the only alternative Gandy considered, although it was the one he and the society favored. In 1936 Gandy compared smokeless zones with further alternatives for smoke abatement, including legislation banning smoke altogether, taxation of coal for domestic use, subsidization of approved smokeless fuels, and licensing the use of coal (Foster, 1972). World War II set back the society's campaign for proposing the zone idea, but after the war the society encour-

aged pilot projects and gained acceptance in two cities, Manchester and Coventry. Then acceptance slowed until new impetus was provided by a crisis event.

As noted in Chapter 3, the December 1952 smog in London led to the Clean Air Act of 1956, which provided for the establishment of smoke-control areas at the discretion of local authorities and with financial support from the central government. The national acceptance of the smokeless zone idea took 20 years from the time it was first advocated. When it was eventually adopted it became an effective tool in smoke reduction. Dramatic reductions in smoke concentrations in the atmosphere of London and other major cities can be attributed in part to the smoke control provisions of the 1956 act. The success of the act, however, and of the smoke-control zone concept owes much to the fact that it coincided with a trend in national energy policy away from coal (Auliciems and Burton, 1973). In an interesting reversal, the idea of the smokeless zone is now being advocated as a model for the further control and abatement of noise.

Maize Breeding

Work in maize breeding did not begin in Kenya in an organized way until 1955, 15–20 years after the development and adoption of hybrid varieties of extraordinary vigor in the United States. Large European-owned farms were producing maize commercially in Kenya in the 1920s. The commercial area required a late-maturing variety (7–11 months) in the interest of increasing profits. It was not until 1957 that work began on early-maturing varieties, where the aim was reversed: to develop a variety that is high-yielding but can complete all essential growth in the 60–65 days of rain during the growing season in the dry Eastern Province of Kenya. By now the early-maturing varieties have evolved serially into a set of crosses of sufficient improvement to be released to farmers. One survey showed 80 percent of farmers using some of the improved variety, but for only 40 percent of their maize crop. By 1974, seed supplies were still irregular, misapplication occurred frequently, and some of the potential yield benefit was not realized because the seed was not accompanied by improved practice. A diagnosis of these problems (Mbithi and Wisner, 1974) indicates a prognosis for a fairly substantial improvement in yield and reliability.

Comprehensive Floodplain Management

The idea of comprehensive floodplain management in the United States developed from research investigations that demonstrated the inadequacies of exclusive reliance on a limited range of adjustments designed to control flood flows through the use of dams, levees, dikes, flood walls, and channel improvements. As a wider range of adjustments was added to the lexicon, there arose the problem of designing an overall strategy of response to flood hazard that would combine adjustments in a more efficacious way. A comprehensive list of adjustments to flood took shape as outlined in Chapter 3.

Combining a selection of alternatives into a comprehensive strategy is by no means a novel procedure: it has been widely used in city and regional planning. The idea of comprehensive river basin development is also closely linked to flood damage problems. The development of a "comprehensive" approach to floodplains can be seen as an expansion of the concept from broader strategic planning into more detailed levels of management.

Hurricane Modification

The practice of modifying clouds by artificial seeding began in 1946 when Schaefer and Langmuir discovered that by seeding super-cooled clouds with freezing nuclei, such as ice or an ice-like substance (e.g., silver iodide), the water droplets will freeze and release latent heat of fusion.

The first hurricane-seeding experiment was conducted by the United States in 1947 under Project Cirrus, although at the time not enough was known about hurricanes to base the experiment upon a theory predicting the outcome, and aircraft research was not far enough advanced to make any before-and-after measurements. The experiment became the subject of considerable controversy, since the seeded hurricane, heading seaward, abruptly reversed its course and moved into Georgia, where it caused considerable damage. It became clear that future experiments would have to be based on a sound foundation of knowledge regarding the internal mechanisms that power a hurricane. This motivated the National Hurricane Research Project in 1955, and the launching of Project Stormfury in 1962, to achieve better understanding of the distinctive processes of hurricanes and how to modify them.

The theory on which hurricane-seeding experiments are now based — the Stormfury hypothesis, as first proposed by Simpson in 1961 — assumes that by seeding the clouds near the area of lowest pressure, the latent heat released will raise the temperature and decrease the pressure in this region, and hence reduce the pressure gradient. The net result of seeding a hurricane would thus be a reduction in the maximum winds, from redistribution of the energy concentrated around the storm center. Over the years the Stormfury hypothesis has been subjected to considerable criticism. Some critics argue that the magnitude of the proposed changes would be undetectable within the background of the hurricane's natural short-range variability, and even that the chain of events would be different from that proposed.

The first full-scale modification experiment was conducted in 1961 on Hurricane Esther; this was followed by the seeding of Hurricanes Beulah in 1963, Debby in 1969, and Ginger in 1971. That the number of seeded hurricanes has been small is the result of certain constraints; thus, in order to be eligible for Stormfury experiments the probability must be less than 10 percent that the hurricane center will come within 80 kilometers (50 miles) of a populated area during the ensuing 18 hours (Howard, Matheson, and North, 1972). Expansion of Project Stormfury into the Pacific Ocean, where the number of tropical cyclonic storms satisfying these conditions is much larger than in the Atlantic, was then considered. Funding was cut back, but the experiment remained a challenge to collective action.

Haphazard Search

Rarely is a systematic search made for all possible adjustments to a hazard. Those considered are usually the traditional ones, plus one or two innovative ideas. Yet the large potential for the creation of new alternatives is demonstrated by the examples cited earlier. The smokeless zone was inspired by the "noiseless zone"; comprehensive floodplain management, by comprehensive river basin develop-ment. The principles of varietal maize breeding were well known but needed application to a newly recognized need for a drought-resistant, high-altitude variety. Cloud seeding for precipitation had been operational for years before the first major hurricane-seeding attempt was made. All the behavioral constituents of floodplain management had their analogues elsewhere—in zoning, maps, insurance—they only required special combination or application.

This is encouraging, in that it suggests considerable improve-ment in hazard adjustment with the application of what is known to be practical. When technological and social fixes are compared, the findings are discouraging. Katumani maize was bulked and made available within 6 years of its development, although its full availability and correct usage require more effort. Hurricane seeding may be adopted prematurely, but there will probably be little reluctance to use it if its efficacy is attested to by responsible scientific opinion. On the other side, the two social inventions — the smokeless zone and comprehensive flood management — took upward of 20 years for authorization by national governments, and then the adoption process remained relatively slow. Some reasons for the preference given to technological adjustments over social fixes become apparent when we consider the way government agencies in both high-income and low-income nations evaluate adjustments.

While enlarging the range of perceived alternatives, group action tends to overlook other possibilities than those selected. As organizations such as ministries and development agencies evolve to handle large-scale technological alternatives, their ability to address themselves to small projects diminishes and a bias emerges in favor of the large and conspicuous undertaking. The long lead time between initiating and completing such projects requires the creation of institutions to develop the plans. Such institutions tend to expand and then after their initial missions are completed, to persist in favoring the alternatives toward which they first addressed themselves. In time they harden into a set of ministries, agencies, bureaus, and state corporations and work out among themselves well-defined and specialized roles.

In consequence, the new alternatives emerging are not noticed when they fall into the interstices of the organizational chart or threaten (by their overlap) the agreed responsibilities of two or more agencies. The capacity to carry out an organized search for alternatives — a major advantage of collective undertakings — is rarely addressed to new areas of research or inquiry which do not fall within the purview of established organizations. Thus, over much of the world, those who drill deep wells for water never discuss them with their counterparts who care for the grasslands surrounding them. Large dam projects are constructed independent of their ecological and human impacts. And those who issue warnings seldom stop to ask, Is anyone listening?

ADOPTION OF ADJUSTMENTS

Collectives that serve as managers adopt adjustments, much as do individuals, for the self-interest of the group constrained by limits of experience, capital, and the perceived efficacy of said adjustments. Those that dispense hazard services do so in response to professional role, to responsibility, or to consumer demand, encouraged or discouraged by need for survival or expansion. The guidance a society exercises on individual or group adoption reflects the prevailing political and social ethos of a society, as well as its organizational ability to handle relatively rare undertakings.

Individual and small-group adoption of hazard adjustments have been studied more extensively than that of collectives. Relatively few formal studies have been conducted, and these largely in North America. Comparative community adoption of drought adjustments for urban water supply has been studied in the state of Massachusetts (Russell et al., 1970), for smoke abatement in the United Kingdom (Foster, 1972), and for floodplain regulation in a few Midwestern communities (Simkowski, 1973; Shaeffer et al., 1967). The sequence of community reconstruction decisions and issues related to future hazard vulnerability was studied for the Managua earthquake of 1972 and the Rapid City, South Dakota, flood of 1972, and retrospectively for the earthquake of San Francisco in 1906 and of Anchorage, Alaska, in 1964 (Haas, Kates, and Bowden, 1977). Extensive studies of organizational change for hazard coping have been made by sociologists (Drabek and Haas, 1974).

From this body of material—large but limited in relevance—four themes particular to collective action emerge to differentiate the description of their adoptive process from that of individuals: (1) the role of crisis, (2) the interaction of role and responsibility within collective groups, (3) the interaction between competitive collectives and their constituencies, and (4) the problem of maintenance or implementation of adjustments when adopted over time.

The Role of Crisis

As noted earlier, the keenness of perception or recognition of a hazard by the general populace often depends upon the recency of a disaster or crisis. Disaster control and relief organizations, the Red Cross, the Small Business Administration, and similar groups that are responsive to expression of public concern tend to reflect this crisis- or disaster-generated pattern, and may be quick to act following a real or imagined disruption. Chief exceptions are the hazards so pervasive that hazard management is part of normal operations; these include fog at airports, where radar landing systems can be adopted, snow that must be removed to keep city arteries open, and drought and its effect on a municipal water-supply system.

The drought adjustments of the water management system are part of the absorptive capacity of the municipality, part of its everyday response to the variability of water supply. In U.S. municipal water systems about $200 million is spent annually on storage facilities to prevent drought (Wollman and Bonem, 1971). In this context it has been suggested that the role of a perceived drought crisis may be to provide support for previously prepared plans or to spark the formulation of additional ones (Russell et al., 1970). But the process of adjustment is an ongoing one, and in Arey's examination of a 100-year record of the financing of water-supply increases, a weak correlation was found with the occurrence of possible precipitation shortfalls. For the nonpervasive hazards, considerable evidence suggests that the perception of these hazards by the public and public bodies comes mainly in the form of disaster and crises, modified by competition for their attention, and reflecting broad cultural change in sensitivity.

Evidence of general public awareness of hazard events is found in articles published in newspapers dealing with air pollution themes. In these the number of articles fluctuates with air pollution crises and is modified by long-term trends in concern for varying social needs and short-term competition for public attention. Thus, the number of articles in U.K. newspapers (see Figure 5.5) is clearly episodic, following major dramatic episodes in 1952 and again in 1962, and the initiatives of government in proposing air pollution control measures. But more recent publication reflects the broad cultural change of increased sensitivity to environmental hazards of all types.

Evidence that there might be only a limited public attention span to crisis comes from U.S. and Canadian data. In these nations major episodes were very much localized: as measured by articles, increase in attention is due to the rise in sensitivity as part of the environmental revolution, whereas decrease reflects the decline in environmental interest after "Earth Day, 1970." The sharp fluctuation in air pollution articles can be explained also by pressure from competitive happenings: the political events and assassinations of 1968 in the case of the United States, the Quebec political kidnappings of October 1970 and the strong government countermeasures in Canada.

Collective action reflects episodic crisis as modified by secular trends in awareness and concern for hazard. National legislation in the United States involving flood exemplifies this. Major legislation in the twentieth century was preceded by great floods, and in only one instance was important action taken without the press of an antecedent crisis. Crisis plays a role in smaller organizations as well; response to disaster has initiated change in a few organizations, but

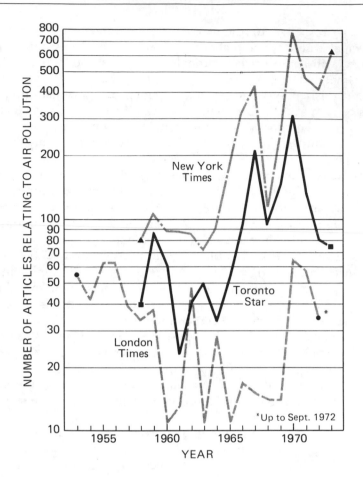

FIGURE 5.5. Air pollution in the media: Canada, United Kingdom, and the United States. Air pollution stories in the media increase with the occurrence of severe episodes (U.K., 1952, 1962) and the general rise in environmental consciousness (post-1965), but decrease with competitive news of political assassinations (U.S., 1968; Canada, 1970).

is more likely to have a lasting effect by accelerating already ongoing changes, as in the construction of municipal works planned already in an area devastated by tornado. Industrial firms reportedly require a crisis to initiate corrective activities (Cyert and March, 1963).

All crises do not have equal weight in stimulating response from groups. A factorial stimulus is visibility, which often appears to diminish with distance from the main centers of influence and power. Thus, the widespread drought of 1952–1956 in the U.S.

Southwest had much less visibility and hence less help from Washington than the drought of 1962–1966 in the more densely settled Northeast. National government action to speed recovery was slower and less effective in the geographically remote area of western Sicily after the 1968 earthquake than in Nicaragua after the 1972 earthquake that occurred in the capital city, Managua. That the earthquake crisis was followed by greater and swifter recovery and restoration at Managua than in western Sicily (Kates, Haas, et al., 1973) may be attributed at least in part to centrality and visibility. The city of Managua is central to Nicaragua and clearly visible from Washington; Sicily is a remote province of Italy, far away from Rome and much less visible in Washington.

Elites, Influentials, and Masses

Understanding of the practical process of collective action is both simplified and complicated by the diversified and specialized roles played by officials within organizations dealing with hazards. With specialization, communication is formalized, more is "written down" as to the hows and whys actions are taken by municipal officials, meteorologists, consulting engineers, and other authorities. But interaction within collectives may also be complex, and individuals within them may behave toward their fellows in ways unrelated to the substantive questions of hazard and hazard adjustments. To illustrate: engineers who serve either as technical personnel or as external advisers have a distinct sense of the authoritative role they should play, and their behavior reflects a concern to safeguard and enlarge their leadership role, independent of the character of the disaster itself.

Participants in collective choices of action may be roughly grouped as the specialized technical elite, the influential decision makers, and the mass publics. Arey has developed a model of such action for one common hazard-related managerial system — municipal water management (Russell et al., 1970).

Participants in the Choice of Municipal Water Supply

In most parts of the world the management of piped water supplies for urban areas is a local responsibility. Among the many duties of the authorities who provide water of sufficient quantity and quality are to anticipate and deal with the risk of drought.

A municipal water management system usually has a three-part division of jurisdiction: elected or appointed community officials with broad responsibility, bureaucratic or departmental personnel with specialized operative responsibility, and external advisers with specialized technical expertise. In Figure 5.6, the municipal water-management system prevalent among cities and towns in the state of Massachusetts is shown. There is no single manager, and even the key groups have diffuse management, with external advice from both public and private sources; officialdom split among the mayor (or city manager), the city council or the selectmen, the finance committee, and the water commission (or board); and operational responsibility separate or lodged in general public-works responsibility. No less important in certain situations may be the users themselves, with both special needs and community concerns.

How do these diverse groups make critical decisions for the system, within the framework of both national policy and the traditions of the community? Arey (Russell et al., 1970) provides a concise description of an extended process:

> The initial impetus for discussion and decision of a system change may come from the elected officials, department personnel, or the water-users. Department personnel are interested in maintaining a relatively adequate system in terms of safety, aesthetic quality, and abundance. . . . They will probably be the source of proposals for expansion in times when the system is apparently performing well, for they have a real interest in insuring continued success and no particular responsibilities for other areas of the municipal budget. Such proposals must be "sold" to the elected officials and their public, and this may be a difficult task in the absence of clear "need." The department may seek out, on its own, the expert opinion of consultants and use this to bolster its case.
>
> The elected officials can hardly be against safe, clean, cheap, and abundant water, but they do have responsibilities in other areas. Faced with chronically insufficient local tax revenues, these officials must balance competing interests. They may agree publicly with a system expansion proposal, while maneuvering in private to delay or kill the measure.
>
> The public, anxious for service, but equally anxious not to see increased debt or taxes, may be reluctant to approve system expansion in normal times. If, however, a drought occurs or if distributional inadequacies are exposed, public concern may align itself with proposals for investment in new sources or transmission facilities. Indeed, a serious enough dry spell may find the public or its elected officials taking the lead in demanding system improvement. It seems often to be the case that drought serves in this respect as a determinant of increment timing, creating public acceptance of previously prepared plans. Moreover, it may spark the formulation of additional plans.

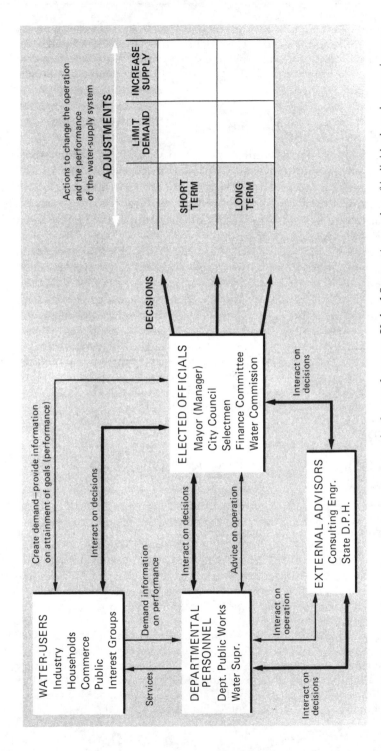

FIGURE 5.6. Municipal water management system in the northeastern United States. A complex of individuals and groups interact to make collective choices of adjustments to cope with drought hazard in the several hundred municipal water systems of Massachusetts.

In the planning process, the scope of the debate is generally set by the consulting engineers. They define the alternatives, make the demand projections, and provide the cost information. They are, of course, constrained by existing attitudes of both officials and public. (pp. 173–175)

The Labor Group for Vegetable Cultivation

The municipal system, while having many participants interrelating in complex ways, usually has distinguishable roles. In the Liu Ling People's Commune, the roles, by intent, are less clear, and at times participants function collectively and simultaneously as technical elite, influentials, and masses. The commune in Yenan derives considerable income from the sale of vegetables requiring frequent irrigation. In the words of the leader of the labor group for vegetable cultivation:

> We are, of course, forced to water our vegetables a lot. We dug the well at the vegetable fields in 1957; that's the best water in the whole district, considerably better than the water here in the village. Vegetables need a lot of water, especially during the dry season. But we had to employ two donkeys and one man every day pumping up the water: one donkey in the morning and one in the evening. Altogether that cost ten yuan a day. Seeing that it was so expensive, we discussed buying an electric pump. In February 1961, the labour brigade decided that we should have an electric pump. (Myrdal, 1963, p. 133)

How did the brigade decide? The bookkeeper reports:

> When we got electricity here, we installed an electric pump. The Old Secretary, Li Yiu-hua, came back from a conference he had been attending. He had seen there that other labour brigades had electric water pumps and electric mills. That had made a profound impression on him. When he got back, he said: "We could have an electric water pump down by the vegetable fields." At the meeting of the representatives on 29 and 30 August 1961 we discussed this idea. We had calculated that the method we were then using to draw water, by donkey, was costing us 240 Y a year for the donkey power and the person in charge of the donkey, and that the actual drawing cost 200 days' work a year. That meant that the total cost of watering was considerably more than 400 Y a year. We had also talked with people from the electricity works in Yenan and they had told us: "If you buy an electric pump, this will involve a big outlay the first year, because the motor is expensive, but after that it will pay you."

We calculated what it would cost to use an electric pump. Purchase price and running costs. We discovered that we should get our money back after little more than a year. If we added to this the fact that we should also have an extra donkey and man to use in the fields, we arrived at the result that it would really involve a gain of almost 400 Y the very first year. Everyone in the brigade committee had been agreed that this was the best and cheapest solution.

The representatives went into the matter very thoroughly. Two of them were strongly against the idea. There were Liu Tsung-hai and Li Wong-shen, both from the village of Wangchiakou. They said: "Who knows if this electricity really works. It sounds all right, but what happens if we're cut off? If it shouldn't work after we've spent so much money on it, we'll burn up with mortification." Wangchiakou hasn't electricity. There are plans to electrify the village in 1965. This has been discussed with people from the electricity works. They have approved the electrification plan. When these two from Wangchiakou had spoken like that against the idea, many others, who before had thought a pump a good idea, also began to hesitate. The representatives were on the point of voting that it shouldn't be bought; but then Fu Hai-tsao said: "I have visited another labour brigade. There they had big vegetable plots and everything was done with electricity. It really is both cheap and efficient." Then Li Yiu-hua told what he had seen and the discussion continued. In the end, the opponents of the electric pump said: "If it is true that it functions, that's all right. We only want to be quite sure before we decide. But if you say that it really is so that it is more efficient and cheaper and sure, we are only pleased. In that case we'll vote for it." After that everyone was agreed, and the proposal was accepted and all the representatives were happy. At the end of the meeting we all forgathered and ate wheat noodles. They bring luck and it was propitious to eat them after making that decision. The noodles were paid for by the welfare fund. This decision that we should eat at the welfare fund's expense was taken last thing at the representative's meeting. (p. 191)

From benefit–cost analysis of noodles and donkeys to that of electricity, the commune representatives range over the many roles of decision making.

Competitors and Constituencies

Successful collective groups often have near-monopolistic roles and responsibilities. Communities have one government, one weather service, one utility; yet even in such situations competitors may arise. Communities have overlapping municipal and regional administrations; specialized weather services exist for air travel and for agriculture; competing resources of gas, coal, oil, or electricity are available to heat homes. Hence it is not surprising that, in hazard adjustment, competitive relations develop around overlaps or vacuums of role and responsibility. For an understanding of why cities and nations may opt for one alternative to the exclusion of others to cope with extreme events, it is helpful to recognize how agencies and their constituencies can shape the choice.

A classic case in point was the struggle (described below) between the U.S. Army Corps of Engineers and the U.S. Department of Agriculture for prime responsibility for flood control. The controversy was rooted in differing scientific explanations of the origin of floods and ways to control them, in the bureaucrats' ambition to maintain and expand their respective organization, and in the personal interests of the urban and rural constituencies with whom they maintained close links.

The Big Dam–Little Dam Controversy

After flooding on the Ohio River in 1907, Gifford Pinchot, crusading American forester and conservationist (and later Governor of Pennsylvania), declared, "The great flood which has wrought devastation and ruin in the Upper Ohio Valley is due fundamentally to the cutting away of the forests on the watersheds of the Allegheny and Monongahela Rivers" (see Leopold and Maddock, 1954). These and similar comments were to set the tone for a scientific debate that persisted long after the immediate issues were resolved. They persist today in the minds of many around the world. The view that deforestation alone causes floods ignores the ample evidence that great floods occurred in pre-Columbian time, when the forests were intact across the American landscape.

By 1936, when the major federal legislation was passed for flood control, the act divided responsibility for flood control investigations between the Department of Agriculture and the U.S. Army Corps of Engineers. So long as the Agriculture people concerned themselves with land-treatment measures—forest and woodland management, strip cropping, contour cultivation, and range management—there was little competition. The effects of such measures on floods were debatable, but they were useful for conservation farming and productivity. It was only when the Department of Agriculture began to build flood-detention works smaller in size but similar in purpose to those of the Army Engineers that the quarrel erupted, as both agencies competed for budget, role, and responsibility.

The issue as to the relative efficacy of a few large dams or many small dams was obscured by persistent conservationist notions that land treatment was sufficient to control floods — or at the very least that land treatment plus small reservoirs that did not substantially flood upstream farmland were sufficient. Thus, rural areas joined with the Department of Agriculture in supporting upstream flood reduction and preserving their farmlands from serving as reservoir sites. Support for large dams came from cites and towns along the mainstream of large rivers subject to recurrent flooding; these constituencies also supported the Corps of Engineers on navigation and water transport projects.

The controversy, which raged publicly in the Congress and in the Executive in the course of budgetary and agency reviews, was in part technical and scientific, and concerned the serious disagreement persisting today as to the role of land treatment in flood prevention. It was also in part bureaucratic, and reflected conflicting aspirations for larger organization and larger budget and the differing constituencies (urban and rural) from which the two "establishments" drew strength and to which they were answerable in both the Congress and the electorate.

The controversy was "resolved" by "compromise" in each of these domains. Gradually the view that large floods cannot be controlled by land treatment and small reservoirs won wider public acceptance, while on their part the proponents of this view conceded that the latter measures might be effective for small floods. At the same time, with increased suburbanization the urban and the rural constituencies began to blur, and the Department of Agriculture began to interest itself in preventing urban flood danger in suburban and rural communities. Finally, a bureaucratic compromise took place — with the help of friends in the Congress — when the two agencies simply divided the flood-control domain into areas of less than and more than 250,000 acres, the Corps of Engineers taking responsibility for the larger drainage areas, and the Department of Agriculture for the smaller.

With this comparison and a gentlemen's agreement not to examine or criticize each other's projects, a great spurt in the building of floodplain works began, to the mutual enhancement — organizationally, at least — of both agencies, if not in the public interest. For while the controversy was alive, it served at the very least to provide some competitive flow of ideas around which differing public-interest groups might gather. The net effect was to reduce the overall commitment to engineering works in the face of controversial and differing claims. With the settlement, it would take another decade for disillusionment with all engineering works big and little to become widespread enough to encourage a broad program of floodplain management.

Public Interest and Special Interest

As the complex of competitive groups expands, so do their relations with special-interest groups. Public agencies, for example, may be responsive to public-interest groups whose influence far exceeds their size. In a study of public decision making in coastal erosion management in five U.S. communities on the Atlantic seaboard, less than 5 percent of the affected public participated personally (Mitchell, 1974). Similar small numbers have been found in the few other environmental hazard studies focusing on public process — studies of drought, air, and water pollution (Kasperson, 1969; Foster, 1972; O'Riordan, 1971b; Burton et al., 1974).

The role of the public has, however, broadened very rapidly in a number of countries. In general, those expanding processes of public participation do not mean that agencies cease to respond to special-interest groups but, rather, that the needs or demands of a wider range of such groups are considered. In contrast, a very different public involvement is characteristic of the People's Republic of China. There, projects to prevent or modify natural events, large and small, are decided upon, planned, and then carried out by very large numbers of individuals and groups without special technical expertise or role responsibility, in accordance with broadly defined national policy. This decentralized integration of public choice, planning, and implementation may be unique in the world.

Differentials in Adoption of Adjustments

The evidence reviewed to explain individual adoption of adjustments related primarily to experience, wealth, the capacity to undertake adjustments, and their importance to livelihood; to a lesser degree, individual variation and personality were considered. Some of the explanations seem to apply to collectives as well, as evidenced in the few comparative studies of adoption that we have. In adoption of smoke abatement in the United Kingdom, communities that adopted smoke control legislation either had suffered recent episodes of serious pollution, or were affluent communities desiring to enhance their environment, or had professional staffs who perceived the severity of the problem and were capable of carrying out the legislation, or had special personalities in the form of interested local leadership. Communities that should have adopted the legislation, by virtue of experience of the real hazard, but did not were poor, mostly coal-mining communities where free coal for the miners was an important addition to income and where smoke control conflicted with livelihood relations (Foster, 1972).

Similar findings regarding experience emerge from the studies, previously quoted, of drought in Massachusetts. In 80 of 82 systems where managers perceived the supply inadequate to cope with the prolonged drought of 1961–1965, restrictions on water use were adopted, whereas they were applied in only 8 of 68 communities whose supplies were perceived as adequate. In 4 of the 5 communities studied by Mitchell (1974) where community coastal erosion adjustments had been adopted, private-interest groups and local officials were very much concerned with soliciting public aid to halt erosion. In these 4 communities the managers had all experienced erosion and were able to estimate losses along the shoreline without significantly differing from the actual recessions.

MAINTENANCE AND CHANGE

A special problem of collective adoption of hazard adjustments is their maintenance. Flood levees may weaken or become eroded because of lack of local maintenance; irrigation canals become clogged or silted; and warning systems atrophy from lack of use. This latter is a special problem. Warning systems and related disaster-preparedness activities lose their state of readiness over time unless frequently activated. How to keep the system in a state of preparedness is a difficult and persistent problem of hazard adjustment.

Despite much study by sociologists of disaster preparedness, there are few findings other than the importance of prior experience and certain organizational capacity for sustained disaster preparedness (Mileti, 1975). Thus, efforts to maintain preparedness should focus on increasing experience: linking responsibilities to groups that have other emergency responsibilities — utilities, police and fire departments, hospitals, armed forces — and thus receiving fairly constant practice in emergency response; grouping responsibilities for multihazards in a given agency, thus providing more practice in meeting these rare and demanding responsibilities; and incorporating some specific emergency preparations in the routine of organization life—for example, taking inventories of supplies and maintaining rosters of emergency personnel.

In the short run the process of collective choice appears erratic; close up, it appears complicated by the unique and situational, haphazard and unpredictable. But in global perspective certain trends are discernible—the mix of constituents changes, the direction does not. The collective function in hazard adjustment is characterized by enlargement of scale of adjustment and by the assignment of specialized roles and responsibilities, hierarchically coordinated and integrated. At any moment of regional and national history, responsibilities for dealing with hazard events are highly intermixed between local communities, corporations or communes, national agencies of government and industry, and private citizens or small groups of private citizens. The broad historical pattern of evolving and shifting public responsibility has an impact on the range of individual choice.

The adoption of a new adjustment at the community level may create new theoretical possibilities at the individual level. It may lead to the abandonment of some individually adopted adjustments. The designation by a community in Britain of a smoke-control area automatically makes available to each householder a subsidy that pays up to 70 percent of the "reasonable costs" of converting his domestic heating system from coal to smokeless fuel. At the same time that these new theoretical possibilities are created, certain individual adjustments to air pollution currently in use may be abandoned. People susceptible to air pollution–related diseases may abandon their use of smog masks, or exercise less caution about their health by smoking more, or engage in more outdoor exercise during pollution episodes, or travel with greater frequency into more highly polluted sections of the city.

The adoption of a new adjustment at the national level may have similar consequences. It may create new theoretical possibilities at the local level for individual and collective decision makers. As these new possibilities are created and adopted, the corresponding abandonment of other adjustments may occur. In the Lehigh Valley in Pennsylvania, the warning systems, floodproofing, and flood-fighting preparedness of the Bethlehem Steel Corporation have been strengthened by a federally built flood-protection levee, the combination thus providing for the community the widest array of flood management alternatives. Nevertheless, the effect of the levee may well be to reduce the disaster preparedness of the corporation by eliminating the frequent lesser floods.

The task of hazard management involves a careful selection of strategies that will combine individual, community, and national action in the most effective way possible. A decision at any one level must be understood in terms of its likely effects at other levels. It is only at the national level, however, that an overall strategy taking cognizance of international implications can be designed.

CHAPTER SIX

National Policy

The belief that national policies are needed to deal with natural hazards in a systematic and consistent fashion has been slow to develop. In the face of many pressing demands, the common response of government has been to provide help only when an overwhelmingly strong case can be made. Most typically, this has meant involvement solely in critical disaster situations, coupled with a slow but crisis-generated expansion of public activities into three policies of hazard control and prevention.

The disastrous Miami River floods of 1913 in Ohio, for example, led to the organization of the Miami Conservancy District, the aim of which was to provide flood protection through reservoir construction. This was undertaken almost entirely at local expense. Similarly, following the major floods of 1874 on the Mississippi, the Mississippi River Commission established to deal with navigation and flood problems constituted a small step in the direction of federal involvement in flood control programs (White, 1945). The dust bowl conditions of the 1930s on the North American plains stimulated new national programs for helping farmers to deal with drought, as illustrated by the measures taken under the Prairie Farm Rehabilitation Act in Canada.

Until 1969, drought relief in Tanzania was a subregional responsibility, and in Australia a state responsibility. In the decade since independence, this evolution took Tanzania toward a policy of disaster relief with a few initiatives for irrigation and development of drought-resistant crop varieties; whereas, in Australia, federal drought policy encourages loss sharing, event control, and damage-potential reduction.

That the emergence of national policies to deal with natural hazards problems has been slow is not necessarily to be deplored. Natural hazards have taken their place in the unfolding order of government responsibilities, and hazard policies are only now beginning to develop in the most affluent nations. Along with social security against illness, loss of employment, and other social stresses, governments increasingly turn their attention to the protection of citizens from freak and unusual events in nature. Protection at all times, however, and for all hazards, remains a remote goal even in the wealthiest of societies. In some instances the time now seems appropriate for the creation of national hazard policies where they do not exist, and for the reshaping of policy where they do.

The choice of a strategy that a nation will adopt in dealing with an extreme geophysical event might logically require, as with both individual and collective action, an appraisal of the magnitude and extent of the hazard itself in a national context, and a canvass of the actions available. A considerable range of choices now exists in theory, and some of these choices may be combined in the design of a national policy.

NATIONAL APPRAISALS OF HAZARD

just as the assessment of all the natural hazards that can occur in a single place (see Chapter 2) might provide a basis for individual and community action, so the measurement of magnitude and extent of natural hazards over a national territory can be helpful in formulation of national policy. Assessment of natural hazards for a national territory tends to be made on an ad hoc hazard-by-hazard basis, as seen, for example, in a recent French mapping of all flood-prone areas (BCEOM, n.d.). Overall national appraisals are, however, very rare.

The Former U.S.S.R.

A recent paper from the then Soviet Union is unusual in recognizing the polygenetic character of hazard events (Gerasimov and Zvonkova, 1974) and describing the national distribution and main characteristics of volcanoes, earthquakes, tsunamis, mud flows, avalanches, floods, hurricanes, drought, and heavy snowfalls. On the basis of an analysis of prevailing combinations of these natural hazards, twenty-nine regions have been identified for the former U.S.S.R. territory (see Figure 6.1). This regional grouping facilitates recognition of the role of man's activity in exacerbating certain hazard conditions. The usual response in the then Soviet Union had been to lean very heavily on the use of engineering and technological measures such as dams, river diversion, and other schemes for

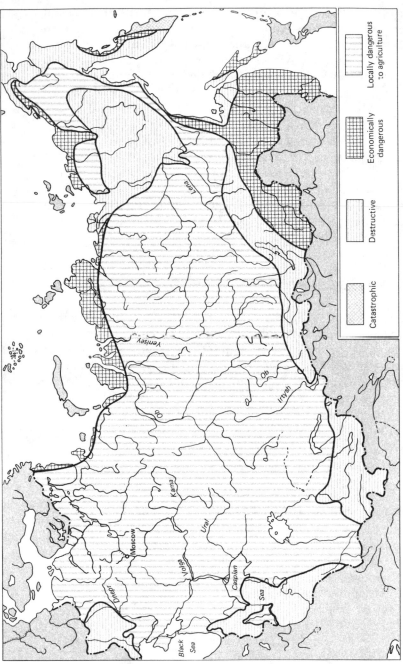

FIGURE 6.1. Hazard regions of the then Soviet Union. Twenty-nine regions have been distinguished in Soviet territory on the basis of prevailing combinations of various kinds of natural hazards. These regions are classified in four groups according to the intensity of events and human impacts. Group 1, catastrophic natural hazards that may cause loss of life and great damage to the economy (volcanism, earthquakes, tsunami); Group 2, destructive natural phenomena that seldom cause loss of life but result in significant damage to the economy, mainly to industry (earthquakes, hurricanes, avalanches, floods combined with other natural processes); Group 3, economically dangerous natural processes (droughts, floods, hurricanes); Group 4, local development of natural hazards that cause damage mainly to agriculture (such as late and early frosts, heavy rains, winds).

166

hazard control, and to provide for compensation of residual damages through the then Soviet insurance agency, Gosstrakh.

Japan

A similar assessment made for Japan (Nakano et al., 1974) describes the high concentration of both population and natural-hazard threats throughout the archipelago. The distribution of population is such that very large numbers of people are exposed to the risk of major disaster arising from earthquake and flood, and this has led to recognition of the need for a wide set of responses. Protective measures are common, especially against floods and tsunamis; and highly developed forecasting and warning systems are now in use.

Canada and the United States

A contrast between neighboring countries is seen in the case of the United States and Canada (Visvader and Burton, 1974). It is not simply that natural disasters are less frequent absolutely in Canada than in the United States; they are also proportionately less frequent when population levels are considered. The population of Canada is about one tenth that of the United States, but the incidence of disaster appears to be a smaller fraction.

This may be explained in part by the geographical distribution of major types of hazard over the North American continent. Tropical cyclones, tornadoes, and drought have their highest concentration and extreme magnitude within U.S. territory. Although the coastal regions of British Columbia and the St. Lawrence River lowlands are earthquake zones comparable to those in adjoining regions of the United States, no earthquakes occurring in Canada approach the damage recorded of those in California and Alaska.

Canada has a flood hazard comparable to that of the United States, but the limited evidence suggests that losses in Canada are probably less than 10 percent as compared to losses in the United States. In the case of the U.S. flood problem, heavy Federal policy since the Flood Control Act of 1936 has encouraged reliance on control technology and has discouraged or given insufficient weight to other possible adjustments such as land-use regulation, flood-proofing, flood warnings, and emergency action to evacuate people and property (White et al., 1958). The extent to which a relatively smaller involvement by the Canadian federal government may have had a favorable effect in keeping losses slightly lower by encouraging use of a broader range of adjustments is not easy to assess.

Some provincial governments have followed a flood policy not unlike that of the United States (Sewell, 1965). On the other hand, the 1959 plan for flood control in Toronto was one of the first anywhere to provide for public purchase of floodplain land (Metropolitan Toronto and Region Conservation Authority, 1959). In both countries the knowledge that a national government can be relied upon to provide postdisaster relief may well have decreased the incentive for individual action and discouraged the search for alternative adjustments.

NATIONAL EXPERIENCE IN COPING

Four patterns of policy are prevalent. In order to illustrate them and to show how they shift over time, the experience with earthquakes in Nicaragua, floods in Romania, and forest fires in North America is reviewed.

Earthquake in Nicaragua: From Disaster Relief to Damage Reduction?

Volcanic eruption and earthquake have intervened in the turbulent history of Nicaragua for more than four centuries. During the last century alone earthquake struck Managua four times: in 1885, 1931, 1968, and 1972. Located as it is along the circum-Pacific ring of seismic activity, Managua is in a classic spot for periodic repetition of seismic and volcanic activity.

In 1931 the population of Managua was about 45,000 people. On 31 March an earthquake struck the city, killing between 1,000 and 2,000 of them and leveling the buildings over an area of 10 square kilometers, leaving 35,000 persons homeless and causing an estimated damage of $15 million (in 1931 values). At this point the options open to the city as a result of the earthquake were rejected, and a decision was somehow arrived at by municipal and national officials to rebuild the city in substantially the same form, in the same location. No new policy initiative was taken, the implicit assumption being that disaster relief operations would continue as and when needed in the future.

Over the next 40 years the city of Managua, which had changed little in the previous 400, grew more than 10 times (see Figure 6.2). The recency of the destructive earthquake and the rapid growth might have provided motivation and opportunity for attempts to minimize disaster impact in some future earthquake, but little was

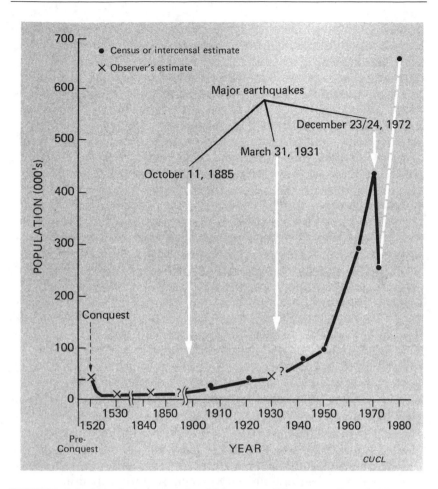

FIGURE 6.2. Population growth of Managua, Nicaragua. It took Managua about 400 years to grow to the size of its preconquest Indian population, and in 40 years it grew to more than 10 times that. At least a third of its population left immediately after the 1972 earthquake, but by 1974 all surviving citizens had returned, or an equivalent number of new migrants had located there, and planners of reconstruction were preparing for a city doubled in population by 1985. Thus, despite more careful construction, more people may be at risk in the reconstructed city.

done. Despite its seismic history and special vulnerability, earthquake disaster prevention or preparedness measures were almost nonexistent. A few buildings (at least six) had been constructed to California seismic-risk standards, and a law providing for installation of seismic resistance in major structures was passed in the early 1970s but appears not to have been implemented. Insurance protection was in force for upper-income housing, since this was required by local mortgage lenders; and a radio frequency was set aside for emergency broadcasts as part of a Central American network programming. To the best of our knowledge, this was the extent of earthquake disaster prevention planning and preparedness in 1972.

The national policy may be characterized as reliance largely on disaster relief, with little or no preventive measures. This was a small step in the direction of hazard control, but the prevailing strategy discounted future earthquake threat. Insurance rates, for example, implied an expectation of earthquake recurrence as rare as 1:100–1,000 years.

On 23 December 1972 three new earthquake shocks produced severe damage in Managua. A magnitude of 5.6 on the Richter scale has been computed for the first and largest of the three shocks. The greatest zone of damage (shaking, faulting, and fire) was in the older, downtown area. Moderate-to-extensive damage, including building collapse, extended everywhere in the vicinity of the city. The earthquake had a shallow focus, such as often intensifies damage; its epicenter was located northeast of the city under Lake Managua.

As dawn broke over the city of Managua on Sunday, 24 December, out of an estimated population of 405,000 at least 1 percent were dead, 4 percent injured, 50 percent of the employed were jobless, 60 percent of all inhabitants were fleeing the city, and 70 percent were homeless. In a nation of some 2 million people, at least 10 percent of its industrial capacity, 50 percent of its commercial property, and 70 percent of its governmental facilities were inoperative. To restore the city would require expenditure of a magnitude equivalent to the entire annual value of Nicaraguan goods and services. With a per capita GNP of about $350 per year, the 75 percent of the Managuan population affected by the earthquake had on the average a loss of property and income equivalent to 3 times that amount ($1,050).

If the seismic past is any guide to the future, Managua will experience future earthquake damage within the lifetime of most of the surviving earthquake victims. Why is this so, and does it necessarily have to be that way?

The record of human settlement provides ample evidence that in areas of seismic risk the risk is often more than balanced by advantages of the site chosen for erecting buildings. Nevertheless, specific patterns of human settlement and activity may pose burdens not reflected in the immediate social benefit to a society. Locational advantage and resource exploitation change through time, and long after these have lost their importance, the organization of a city can take on an existence of its own. Its infrastructure, its everyday activity, and its residents' attachment to it create an enormous inertial mass that persists and often expands.

One year after the Managua earthquake, relocation of the city was no longer seriously considered as a possibility. The pattern of reconstruction was still in doubt. It was not likely that Managua would simply duplicate the past as it did in 1931, but it was not clear to what degree future risk would be reduced through slowing down the growth of population by regional decentralization, through spreading out the city to reduce density, and through improvement in seismic resistance of rebuilt and new structures. Some progress had been made in all these directions. It was countered by the rapid return of the refugee population, pressures for laissez-faire speculation and reconstruction, difficulties in enforcing construction standards, the absence of critical information, and a general atmosphere of indecision and confusion.

In 1973, Nicaragua stood again at a choice-point in national policy. At the least it could continue to pursue the disaster relief strategy drawing heavily in the future, as on this occasion, on the international community for help. On the other hand, Nicaragua could slowly pursue the limited opportunities of hazard control and disaster preparedness, particularly in relation to the secondary effects of fire. It might even use the occasion, with appropriate international assistance, to move quickly into a comprehensive policy of disaster prevention. By 1974 it seemed that some changes in earthquake policy might be instituted but would be few in number. The dominant strategy in Nicaragua was to rely on the approach used in the past—namely, to hope that disaster would not occur and only slightly to prepare to meet it when it inevitably would.

Flood in Romania: From Hazard Control to Comprehensive Damage Reduction

Historically, floods have disrupted communications and commerce and caused great damage in Romania. In 1970 the nation suffered flooding that affected large areas of the nation, with total damage estimated at more than $555 million. One of the more severely affected regions was the Mures River Basin, where more than one-half of the loss was concentrated.

The Mures River watershed is located in the west-central part of the country. The drainage area within Romania is 27,920 square kilometers, but the excessive loss was in the upper portion, which is 350 kilometers in length and 17,000 square kilometers in area. A populace of 1.4 million dwells in this Transylvania Plateau region, with the igneous chain of the Carpathian Mountains rising around most of it to elevations of 1,900–2,500 meters. In addition to flood hazard, other problems involve water supply, water quality, power, soil erosion, and landslides.

Previous national policy placed heavy reliance on flood-control structures in an attempt to prevent the inundation of cities and agricultural land. The 1970 flood disasters, especially those in the Upper Mures, motivated the Romanian government to initiate a comprehensive approach to the problem. Three fundamental aspects are considered in this program:

1. Provision of a system of water development and management intended to reduce the harmful effects of floods and pollution, and to conserve and make more efficient use of water resources. This includes plans for flood control, floodplain regulations, and other nonstructural measures, the provision of irrigation water supply, municipal and industrial water supply, hydroelectric power, water quality, and water-based recreation, and the development of a system for water resource management.
2. Development of a systematic pattern of land-resource use designed to ensure best utilization of the resource and protect it from deterioration.
3. Development and use of land and water resources in coordination with one another and in accordance with prospective social and economic development of the region, so that land and water resources make an optimal contribution to the social and economic development of the region.

In the classic pattern of disaster-relief policies, the Romanian government energetically rehabilitated the areas stricken by floods in 1970. By 1974 it was beginning to undertake a multipurpose program for the overall protection and development of the basin. During the next 5–10 years an estimated $185 million will be needed for structures such as reservoirs and dikes, and erosion control, irrigation, and drainage works. Total investment will be even greater, as it will include studies, nonstructural actions, training, and educational programs. In short, Romania took the occasion of its recent disaster experience to move in the direction of a comprehensive policy for flood-loss reduction and associated activities in water management.

Fire in North America: From Event Control to Comprehensive Hazard Management

In the hunting-gathering cultures that first flourished in the shifting cultivation of tropical savanna and forest, intentional fire was the dominant adjustment. Campfires served the important purposes of cooking, warming, preparing tools, protection from animals and insects, and signals (Lutz, 1959; see also Table 6.1 below). When fires escape through carelessness, they are not necessarily deleterious to man or the natural ecosystem, though they may trigger a

TABLE 6.1. Combinations of Adjustments to Fire under Four Policies

Policy and combinations	Adjustments praticed	Major consequences	
		Natural	Human
Intentional fire	Set fires	Maintenance of plant community	Increased production
Uncontrolled fire	Set fires Localized control	Clear forests Disturb sod, water, animal systems	Rapid expansion of production Air pollution
Uncontrolled fire transition	Relief for sufferers Fire control Education Cultivation practices Mechanical, agronomic	Catastrophic fire Ecosystem instability	Reduced timber losses Enhanced recreation Threatened residential
Comprehensive program for managed fire	Fire control Education Cultivation practices +prescribed burning +chemical treatment Lightning suppression Research programs Genetic selection, etc.	Ecosystem stability	Increased timber production Modified recreation Air pollution

conflagration in areas where there is no lightning to start a fire. In any case, aboriginal man as well as present-day hunters and cultivators have had cogent reasons for widespread burning (Stewart, 1956). It aids in hunting by driving game; in preparing land for cultivation by reducing brush, weeds, and pests; in temporarily improving soil fertility; in increasing yields of wild seeds and berries; and it is thought to improve pasture. Whether or not it was seen as a means of averting catastrophe is not clear.

The question of precisely how influential people have been, in contrast to lightning, in establishing and maintaining those prime areas of grassland or woodland such as that of the longleaf pine continues to excite study and debate (Sauer, 1944, 1950; Stewart, 1963). It would be a mistake to dismiss intentional burning, either past or present, as all bad for natural and human systems. There is ground for thinking that fire-dependent plant communities burn more readily because natural selection has favored their flammable properties (Mutch, 1970), and that people may have contributed to that selection.

A second contribution to adjustment is made when man, in expanding his agricultural land use, employs fire to clear the forest and extend cultivation or grazing operations. Unlike intentional fire activity, uncontrolled fire that is set and allowed to spread for these purposes is not geared to maintenance of the plant community. Such fire requires localized control to prevent damage to settlement, and it causes pronounced disturbance in the ecosystem while promoting expansion of agricultural production. It is likely to be a transitional combination: either the disturbances prove crippling (as when shifting agriculture, excessively intensified, leads to soil destruction), or a forest stand is eliminated. In the earliest days of forest-fire control in the United States, a basic obstacle to observing the ordinance calling for "the utmost tact and vigilance" was resistance of "settlers . . . accustomed to use fire in clearing lands" (Pinchot, 1947, pp. 276–278).

The best recognized combination of adjustment is that developed by government programs for fire control, fire-control operations, education, regulation to prevent man-made fires, and design of agronomic and mechanical forest cultivation practices to reduce loss and vulnerability from fires. Certain results of these activities have been measured, but the record does not permit evaluation of other respects.

In Canada the average number of forest fires reported annually between 1932 and 1967 decreased until the mid-1950s, and increased thereafter with the inclusion of Yukon and Northwest

Territories. In the United States during the same period the number fluctuated more, with an upward trend after the mid-1950s. The total area in hectares burned annually in Canada remained in the 0.6–1.4 million range, while in the United States it declined gradually through the 1960s from a peak in the late 1920s.

By the mid-1960s the area under some kind of government program and protection in each country was some 2.5 billion hectares. The total number of fires, though fluctuating, increased somewhat; the size of average burn had decreased greatly in the United States; the unit cost of fire fighting had increased; and government expenditure for protection showed signs of leveling off. Yet, the possibility of major forest conflagrations remained. This was a crossroads: the government could persist in emphasizing fire control, or it could adopt a more comprehensive policy. Fiscal exigency reinforced scientific argument.

In a variety of ways the combination of adjustments described as fire control had been changing toward a strategy that might be designated as "managed fire." One crucial shift was in the policy with respect to prescribed burning as a principal element in cultivation practices. Another was the increased attention given to systematic scientific investigation of new techniques and of human response to them (Barrows, 1971). Prevention is now considered in all its aspects (Hall, 1969). The research initiatives include studies of lightning processes and of possible suppression techniques, chemical methods of treating fuel sources, attitudes and behavior of people using forests, and insurance — specifically the reasons it had not been used to both distribute loss to and evaluate management measures (Greene, 1965).

This broadening of concern beyond the watershed, range, and economic studies of an earlier time moved toward a more comprehensive and at the same time more troublesome approach, one requiring consideration of a large number of alternatives, as well as assessment of widely diffused effects. While the actual consequences of this combination of adjustments cannot be estimated with confidence at this time (see Table 6.1), it can be expected to strengthen the stability of natural systems, and in the long run to enhance the timber production. It is likely to change the conditions of outdoor recreation, alter wildlife population densities, and increase air pollution over large areas. The wise management of fire thus requires innovative assessment of the whole range of possible adjustments and continued research to provide productive ways of coping with this equivocal hazard.

TYPES OF NATIONAL POLICIES

Four distinct patterns emerge from the welter of activity, and in many cases inactivity, of which Nicaragua, Romania, the United States, and Canada are examples. With much overlap, national policies seem to focus on

- disaster relief
- control of natural events
- comprehensive reduction of damage potential
- combined multihazard management.

The four patterns, as shown in Table 6.2, are not mutually exclusive. Policies of one set may include elements of the other sets: in the example of Australian drought, reduction of damage potential will include elements of the control of hazard-engendering natural events and disaster relief. It is useful, however, to characterize these types and describe examples as a means of providing a framework for thinking about the selection of national policies and strategies.

Disaster Relief

Countries where major disasters do not occur or have not recently occurred feel no pressure toward national governmental action. In

TABLE 6.2. Types of National Hazard Policy

Type	Hazard	Country
Disaster relief	Earthquake	Australia
	Earthquake	Nicaragua (Managua)
	Drought	Tanzania
	Air pollution	India
	Flood	Canada
Event control	Drought	United Kingdom
	Drought	Australia
	Air pollution	Mexico
	Urban snow	Canada
	Flood	Hungary
	Flood	Sri Lanka
	Flood	Romania
	Cyclones	United States
	Cyclones	Bangladesh
Comprehensive damage reduction	Flood	United States
	Drought	Israel
	Air pollution	United Kingdom
Combined multi-hazard management	—	—

Australia the risk of severe earthquake is very low (absolutely as well as relatively), and hence there is no evident need for a national earthquake program. Should an earthquake of sufficient magnitude occur to cause severe damage, the commonwealth government would probably provide disaster relief and take other emergency action. The "Australian earthquake policy" can thus be described as very limited, and probably not one that has received much serious attention.

Canada's flood policy, like Australia's earthquake policy, is a limited one, but not for the same reason. There have been a number of disastrous floods in Canada, among them on the Fraser River (Vancouver, B.C.) in 1948, on the Red River (Winnipeg, Man.) in 1950, and on the Don River (Toronto) in 1954. The federal government helped with emergency relief and rehabilitation, and was inspired by these disasters to assist in the construction of flood-control works — the Red River floodway in Manitoba and the Toronto flood-control and recreation scheme (Burton, 1965). No national flood program emerged, however. There is still no national collection of flood-loss data and no national scheme for flood insurance. Reviews made occasionally by federal agencies indicate a conscious canvass of the situation at the national level and probably a conscious choice to take little or no further action. The preferred national stance has been to regard flood problems as largely a provincial government responsibility, the national government being prepared to assist in the event of disaster of major or "national" proportions and with cost sharing of flood-control works. The Canadian flood policy is thus in transition from a disaster-relief to a hazard-event control policy, although the control works will probably involve little participation by national government agencies. Attempts are now being made at the provincial level to develop a more comprehensive set of loss reduction policies with national coordination.

Air pollution reaches high levels in some Indian cities. In Calcutta the rapid radiation cooling that occurs after sunset and the cool winter season produce temperature inversions that help to create near ground level a heavy concentration of emissions from cars, factories, railways, and the innumerable small dung-fuel fires used for cooking the evening meal and for providing warmth. The resultant "smung" is offensive and hazardous to health. No significant efforts to deal with the problem have been put forth at the national level. In the hierarchy of problems found in Calcutta and in the nation as a whole, the issue of air pollution, though recognized, is regarded by responsible public officials as of relatively low priority. Were an air pollution disaster to occur, relief could be provided, if deemed appropriate, by the state or national government.

No major drought disaster or crisis had occurred in recent decades in the United Kingdom until the 1976 dry spell. The country is relatively well watered. Evapotranspiration is low and rainfall is evenly distributed throughout the year. There was no dramatic evidence of need for a national drought-alleviation or drought-control program. By the 1970s, however, it was apparent that such a need would arise in the future, and there had already been a number of scientific investigations of drought potentialities (from Brooks and Glasspoole, 1928, to More, 1964).

As the volume of water use in both industry and the home increases, it is necessary to construct more water supply engineering works (e.g., storage reservoirs, pipelines, and deep wells) to keep available supply ahead of demand. By the 1970s, projects for water storage and interbasin transfer had been constructed in Wales, and in north Derbyshire and the Lake District of England. These and other developments, such as groundwater extraction in the south-eastern region, were not, however, sufficient to prevent the severe repercussions of the drought of 1976.

Although the possibility of water shortages had been recognized in Britain, for a variety of reasons it proved difficult to develop additional supplies as insurance against severe drought. Among these reasons were financial stringency and the mounting opposition of environmental groups to new reservoir construction, either on land, or, as was proposed, in the estuary areas of Morecombe Bay and the Wash. Another reason was the belief that in circumstances of financial constraints and environmental opposition it was acceptable to take some risk in a generally very well watered country such as the United Kingdom.

When drought occurred in the summer of 1976 a new national policy was quickly developed. Initially this involved the appointment of a Minister for Drought with strong emergency powers granted by a rapidly formulated act (The Drought Act, 1976. Elizabeth II., Ch. 44). Under this act, 150 orders were made within a few months, 110 of them for augmentation of resources and 40 for limitations on water use.

The policy adopted as a result of the 1976 drought was initially a drought relief and emergency provision affair. In the longer run, the drought seems likely to bring about enactment of new policies for water supply. Steps to conserve available resources and to constrain existing use have been resisted in the past; for example, proposals to introduce water metering for domestic consumers have been consistently rejected. Such prudential measures are now likely to be more carefully and more receptively considered.

In situations where a national program is lacking for a specific hazard, the inchoate policy is limited to response to crisis or disaster. Thus, insofar as earthquakes are not likely in Australia, a national program may never be indicated. As India becomes more industrialized and urbanized, however, as the level of national wealth increases, and as other problems are successfully resolved, air pollution seems likely to rise gradually in importance until it becomes recognized as a national issue. The water situation in Britain has already reached the point where, as a result of drought, a significant shift in water resources development policy seems likely. Canada may be said to have moved already beyond a program of flood-disaster relief into a policy of natural-hazard event control, with the major responsibility resting with governmental authorities below the national level (i.e., provincial or metropolitan).

For an industrialized country, disaster relief should probably move toward a comprehensive natural-hazard insurance plan. Certainly in the more affluent nations the tendency to guarantee a minimum level of social security and welfare might be extended to include compensation for loss from natural hazards. Such a provision would have to be carefully designed to avoid creating higher loss potential by encouraging people to take greater risk in the knowledge that their loss would be covered by insurance subsidy. It would also help reduce the further decline or atrophy of community and individual adjustments. This would probably require the merging of the prospective insurance scheme with land-use planning and regulation, and involve premium scales designed to promote individual or community preparedness. For special-risk groups — the indigent, the aged, the propertyless — specially tailored welfare services could be delivered.

In developing nations the reliance on disaster relief raises questions of effectiveness and equity, and the relief effort is becoming increasingly sophisticated in this direction. In some instances, serious attention must be given to controlling secondary effects such as inflation and profiteering. A glaring inequity is created by disasters in which those who are hit not quite hard enough to be considered victims eligible for relief do nevertheless suffer significant loss. Sharply rising prices, exorbitant interest rates on loans and advances, hoarding for profit, and other opportunistic practices can make destitute those on the periphery of a disaster area who may not be eligible for national relief. To supplement traditional market mechanisms, rural credits, price controls on building materials, and similar measures could all help alleviate the plight of "hidden" victims of natural disasters.

Control of Natural Events and Their Effects

A limited range of technologically oriented adjustments for controlling natural events and their consequences is a common hazard policy in industrial nations and countries with a mixed society. This policy begins at the point when a hazard is recognized as requiring more attention than relief. In Bangladesh the national response to cyclone hazard has been restricted largely to construction of protective dikes and embankments (Islam, 1974) and to increased use of warnings on the basis of meteorological forecasts. Other feasible actions, such as emergency evacuation and mass shelter, were not adopted until recently. The many small-scale individual adjustments used by farmers in the hazard zone were not encouraged or supported by national action. The November 1970 cyclones revealed a serious lack of coordination between warnings and individual reactions.

Similar patterns of national response occur almost universally. On the Tisza River in Hungary, flood-control and river-training works were constructed at heavy cost, whereas other measures such as insurance, land-use regulations, construction of buildings on piles or stilts above flood level and agricultural practices such as later or earlier planting, choice of crops, and substitution of livestock for unmovable crops received little attention at the national level until recent moves toward comprehensive damage reduction. In Canada municipal expenditures to deal with urban snow hazard are limited almost entirely to snow ploughing, snow removal, and the salting and sanding of highways. Slight effort is made to explore other ways of reducing the impact of blizzards, and no other adjustments, such as retiming or rescheduling of events or the use of other forms of transportation, are systematically canvassed at national or provincial levels. In Bangladesh, the protective works, however limited, induced people to settle in more hazardous sites. The Tisza River levees when overtopped prolonged the duration of inundation. The abundant salting of Canadian highways pollutes groundwater, kills trees, and accelerates the corrosion of automobile bodies.

A characteristic of control policies is their excessive reliance on technology or brute-force resistance to natural events. Often they prove ineffective in the face of highly infrequent events; they may exacerbate the problems they are designed to alleviate; and they may provoke unanticipated side effects—such as the exposure of more people to these highly infrequent events. One common reason for the ineffectual behavior is that they fail to take individual or community reactions into account or accurately predict counterproductive measures that will be adopted by those who come to realize that government has assumed responsibility to some degree for their protection from natural hazards. Control policies may thus unintentionally increase the damage potential or lead to abandonment of traditional adjustments—folk or individual—previously practiced.

Comprehensive Damage Reduction

As the weaknesses of hazard-event control programs become apparent, there is growing recognition of the need for a more comprehensive canvass and adoption of possible adjustments. Examples of a well-developed comprehensive policy are not easy to find. Perhaps the most advanced is the statement "Unified National Program for Managing Flood Losses in the United States" (U.S., 1966), as noted in Chapters 3 and 5. In addition to the traditional role of federal government in constructing, directly or indirectly, massive engineering works for flood control, this plan as supplemented in 1976 for floodplain management provides for

1. a program to delimit major flood areas and to provide flood hazard information in the form of maps, profiles, charts, tabulations, graphs, and narrative description;
2. development of a uniform technique for determining flood frequency;
3. a new program for collecting more useful flood-damage data, including research on floodplain occupancy and urban hydrology;
4. encouragement and technical assistance to state agencies to use floodplain regulation;
5. use of existing federal services to ensure that state and local planning takes proper and consistent account of flood hazard;
6. action to support consideration of relocation and flood-proofing as alternatives to repetitive reconstruction;
7. an executive order directing federal agencies to consider flood hazard in locating new Federal installations and in disposing of Federal land;

8. programs to prepare and disseminate information, and to provide limited assistance and advice on alternative methods of reducing flood losses, including land regulation and floodproofing;

9. an improved system for flood forecasting;

10. a study of the feasibility of flood insurance;

11. new arrangements for flood-control projects, surveys, cost sharing, and reporting of project benefits;

12. the use of land acquisition by federal agencies and broader loan authority for local contributions to flood-control projects.

The Federal system of flood insurance, instituted in 1968 in the wake of the Unified National Program, made local land-use regulation in floodplains a requirement for qualification for such protection. In 1973 the relief and rehabilitation policy was modified to withhold special benefits from floodplain occupants who had not met the conditions for purchase of insurance and to make purchase a condition of participation in other federal programs such as mortgage insurance. These actions demonstrate the way in which a comprehensive range of alternatives can be developed at the national level, how they require state and local response, and how they are designed to increase the range and number of adjustments adopted at the community and individual levels. In 1992, a further revision of policy was under way.

In Israel the shortage of water and the ever-present threat of drought were followed by development of a water-management policy that seeks maximum use of available supplies. Construction of the National Water Carrier to take water from the north and supply communities along the way as far south as Beersheba is an important component of the plan. Conservation of water in agriculture is widely practiced, extending even to reconstruction of sections of the ancient Nabataean system to see what might be learned from the past. Water conservation in urban and industrial use has developed more slowly, while moving steadily in the direction of a comprehensive policy. Metering is mandatory, production of water-saving appliances is underway, and industries are regulated by a system of allotment allowances. Prices of water recently were increased to a point equaling the average historical cost of providing water, if not the (more desirable) price setting to equal the cost of the most recent supply. Research looks toward reclamation of sewage and desalinization for future supply. Wider adoption of comprehensive national policies for single salient hazards is the most likely emphasis for the next decade.

Combined Multihazard Management

Evolution toward multihazard management involves equalization of risks in the industrial countries and the selection of risks for attention in the developing countries. Recent studies in North America and Europe show wide diversity in the acceptable risk levels of various hazards of natural, industrial, or social origin. The monies expended to reduce or to protect against them, the administrative organization, the research conducted—all vary by great orders of magnitude. Risk management may not call for equalization—the total cost may be excessive and the effort might be futile—but since many such choices as to preference for risk abatement arise by historical accident, the comparative evaluation of such risk levels can help bring balance into the overall effort.

The reverse is suitable for developing countries, where hazard management calls for selective choice, a minor program of disaster relief and prevention for most hazards, and selective efforts related to programs of social or economic development. This linkage of hazard management to socioeconomic development in these countries should be paralleled by linkage to environmental protection and preservation in industrialized countries. In both cases the argument for linkage is persuasive: extreme events cause serious loss of 1 to 3 percent of gross national product, especially in developing countries. Hazard-prone areas often overlap with areas of access to environmental amenity in industrial countries.

Could there be a set of comprehensive programs designed to deal consistently with all hazards faced in a particular country? Such programs would, in effect, combine damage-reduction measures for all major hazards, drawing on a wide range of adjustments and making them interact in a harmonious fashion, with efficient and equitable disaster relief for rare or minor hazards. No such combined program exists. Indeed, there are few enough comprehensive damage-reduction programs for single hazards. As new comprehensive programs are added, they should be designed with a view to fitting them into a combined program. The first steps in that direction were taken in the United States in 1971 in the National Review of Disaster Preparedness made by the Office of Emergency Preparedness (U.S., OEP, 1972), and in 1972–1974 by the Assessment of Research on Natural Hazards (White and Haas, 1975). Both dealt with the full array of geophysical hazards.

DIRECTIONS FOR NATIONAL POLICY

One favorable outcome of the national policies adopted by wealthier nations has been a reduction in the loss of life, physical injury, deprivation, and hardship associated with extreme events. On the other hand, the scale of environmental disruption and ecological damage is large and the actual level of damage in monetary terms tends to increase. While a greater degree of security has been provided, it is not complete safety. Major disasters still occur, exacerbated in some instances by the failure of technology itself, as exemplified in the collapse of dams such as the Viaont Dam failure in Italy in 1963 and the more recent dam failures in Buffalo Creek, West Virginia, in February 1972, at Rapid City, South Dakota, in 1972, and the Teton Dam in Idaho in 1976.

The modernizing forces of folk or traditional societies, especially in the urban-industrial and commercial-agricultural sectors of developing countries, are already moving rapidly along the path cleared by the industrial nations at an earlier date. In the Third World of today, however, is it possible to avoid the fallacies of an event-control type of program and to move directly from a disaster-relief program to a comprehensive damage-reduction program eventually combined into a multihazards policy?

The folk or traditional sectors in Third World countries have an opportunity to make a remarkable transition, and one that is not so unlikely as may at first appear. Comprehensive programs for damage reduction involve the blending of event control and disaster relief with a wide variety of individual and folk adjustments. The task of adding event control and damage reduction programs into a folk pattern of adjustments, without destroying the latter, is probably no more complex or difficult than the task of revitalizing individual responses in an industrial culture. The two efforts can be mutually helpful. A shift may occur from one type of hazard policy to another, particularly under the stress of a major disaster event. When a disaster occurs in a developing country, question is immediately raised as to whether disaster relief should be supplemented by a new hazard policy. It is commonly assumed that a change will be in the direction of an event-control program. Under exceptional circumstances it may be possible for a developing country to move directly from a disaster-relief strategy to a comprehensive approach to damage reduction. In an industrialized country, disaster may reassert and strengthen support for an event-control strategy. It might, however, provide an opportunity to move toward a more comprehensive type of program.

It is not immediately apparent how highly developed or wealthy a society should be before embarking on a national program of hazard management. Nicaragua, with a per capita income of $350, Romania, with a per capita income of $1,000, and the United States, with a per capita income of $6,200 all face critical choices of policy. The desirable character of a national policy depends upon the resources of the government and the nature and degree of the problems involved. There can be no universal blueprint for hazard management. Very probably some countries should have a hazard policy which specifies no program other than disaster relief, with modest attempts at mitigation and preparedness. For others a minimum program might prevent or discourage expansion of hazard vulnerability without diverting major resources to the reduction of existing risks. In others a comprehensive hazards policy might be clearly desirable and could be expected to result in substantial benefits in terms of reduced loss potential. In this effort international collaboration can speed the pace of action.

CHAPTER SEVEN

International Action

To focus on disaster relief in dealing with natural hazards is like expecting the Red Cross to stop a war. While humanitarian concern for relieving suffering in time of crisis deserves encouragement, its effects are largely palliative, and it has limitations for international action toward natural hazards. Nevertheless, disaster relief is the one way of responding to such hazards in which all nations have thus far found it convenient to join, and a Disaster Relief Office was launched by the United Nations General Assembly in 1972 without a dissenting vote. Although the UNDRO is likely to remain for some time the principal agency of international cooperation in coping with hazards, it ought to be regarded as only one avenue for cooperative action among nations.

In view of the fact that no country has a comprehensive, integrated program meeting the criteria suggested in Chapter 6 for managing environmental risk, the agenda for international action are challenging. They range from scrupulous monitoring of extremes in the global environment to scientific investigation, disaster-relief planning and reconstruction, and damage insurance. One or another of these needed programs embraces the interests of every nation, even though some concentrate on regional groups or on developing countries.

As the development of transport and communication brings the peoples of the earth into rapid and more frequent contact with each other, and as the economic integration of the globe makes nations more interdependent, new opportunities for collaboration are created and, ironically, certain of the risks increase. While it is increasingly possible that damage to a specialized industry in one country will impede a user of its products in another country, the possibility also increases that the nation affected may join with fellow nations to prevent such disruption.

The distribution of wealth, of scientific and technical knowledge, and of capacity for emergency response among nations is so uneven that the better endowed have a moral obligation to assist others when that assistance is needed and requested. A further reason for collaboration arises from the temporal nature of the hazards themselves. Because events on the higher end of the magnitude scale are exceedingly rare, the accumulation of experience can be enhanced if nations share their experience rather than continuing to rely on their knowledge of rare events occurring within their own territories.

International concern with the quality of the global environment marked the U.N. Conference on the Human Environment (Stockholm, 1972) and led to much discussion and some active exploration of how nations might better cooperate to preserve the biosphere. Such cooperation is difficult to arrange, in part because of great differences in distribution of political power and material wealth. Poorer nations resent suggestions that they should accept any curbs on the costs involved in their urgent rush toward development and modernization, toward their attempt to provide their populations with minimum goods and services essential for the support of human life with dignity. They are especially unlikely to accept additional restraints in the interest of protecting the environment of a global society in which inequity and injustice are so prevalent and where high-income nations have gained material comfort at the cost of degrading their own habitats as well as those of supplier nations.

Collaboration in dealing with natural hazard is less difficult than maintaining the quality of the environment. Natural hazards are seen as disadvantages and obstacles to development in both rich and poor nations, even though the long-term destruction of the environment from such hazards is not recognized so clearly. The instinct to avoid the fatalities and the property damage inflicted by extreme geophysical events often transcends the economic and political divisions of mankind in the immediate aftermath of disaster. The record is not entirely altruistic, however. As with individuals or regions, nations profit from the woes of other nations. Canadians or Argentinians rich in grain may profit when the droughts and floods occurring elsewhere help to raise world prices. The U.S. government, generally magnanimous, failed to offer aid to an adversary, Cuba, at the time of a severe hurricane, although an American Quaker group did so. In general, however, there is hope that international collaboration in dealing with natural hazard may increase more harmoniously than certain other attempts (the U.N. Conference on Trade and Development, for one) to cope with truly troublesome world resource problems, thus furnishing an example worthy of imitation in the exertions against such problems.

WHAT IS WORTH SHARING?

Certain elementary notions about hazards need to be understood by people responsible for international initiatives. The range of possible adjustments should be recognized, and the potentialities and limitations of methods of investigating them should be known. The present state of knowledge is summarized in terms of concepts, adjustment techniques, and study methods.

Basic Concepts

In the fashioning of national and international programs, three simple concepts stressed in the preceding chapters are repeatedly neglected, with the result that reasoned choice of suitable policies, whether for relief, controlling events, or reducing damage on a comprehensive scale, are impeded. Since they have already been described at length, it will be enough to state each briefly.

1. Hazards always result from interaction of physical and human systems. To treat them as though they were wholly climatic or geologic or political or economic is to risk omission of components that must be taken into account if sound solutions for them are to be found.
2. The use of resources of a hazardous area almost always leads to social benefits as well as social costs. It is essential to identify the trade-offs between the benefits and costs in the broadest sense.
3. In only extremely rare circumstances is there solely one adjustment that merits adoption. Usually there are alternatives that may be as effective as or more viable than the conventional technique or the one that lends itself most easily to public action.

Were these notions embedded in the views of well-intentioned public servants, there would be far less relief that worsens the food production capacity of its beneficiary, fewer flood-control dams that stimulate higher flood damage, more land development that yields net benefit rather than net loss.

Adjustment Techniques

In addition to being aware that there is usually a range of possible adjustment techniques, it is important to know the potentialities and handicaps of each. The potential capacity of each to reduce loss of life and property gives one measure of its possible utility (as shown in Figure 7.1), but must be accompanied by estimates of other benefits it might confer as well as of the cost of applying the adjustment and of thereby forgoing resource use in the area or elsewhere. The estimates given in Figure 7.1 are for all types of situations. Not all could be combined for a single situation. They assume, further, that the adjustment will not be counterproductive—a frequent risk.

Relief and Rehabilitation

When disaster strikes in the form of sudden earthquake, overbank stream, or withering zenith of drought, the key to effective relief

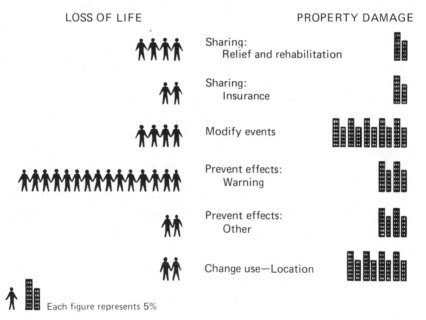

FIGURE 7.1. Loss reduction from international sharing. Comparative maximum potential loss reduction from widespread adoption of current knowledge of adjustments on a global basis. In general, prospects, for reduction of loss-of-life exceed that of property. This estimate assumes a careful and comprehensive application of adjustments. As the effects overlap, the total reduction of such a program is less than the sum of individual adjustments with perhaps 80–85% of life loss and 40–50% of property damage avoidable in total.

and rehabilitation in agricultural Bangladesh or the industrial United States is organizational strength and flexibility. At all levels of government, this requires either an existing agency with other duties, such as public works, capable of responding quickly and surely to the unforeseen, or a new agency prepared to move decisively when the need arises. If there is no preparation, and no agency specially poised to act, an agency trained for a wide variety of social emergencies—such as a national army or a municipal police force, for example—may deliver the most prompt and efficient help.

Such a standby force may be more useful in coping immediately with disaster than an incompetent agency designated to handle emergencies. One confident and tightly organized agency without preparation may be far more effective than a large array of agencies of good will and various degrees of training acting without coordination. At local, national, and international levels the twin need is for preparation and coordination. With such organization, public agencies can marshal the highest skills and sympathies of their citizens to relieve immediate distress and begin the more protracted and intricate task of rebuilding.

Long-term rehabilitation, in turn, is strongly influenced by the degree of preparation and requires different skills, resources, and organization than emergency relief. Unless a fairly suitable plan for reconstruction is already in hand, the pressure (as at Managua) is to rebuild on the old plan of construction, land use, and social institutions. Lack of preparation makes for unimaginative and long, drawn-out reconstruction (see Figure 7.2), as well as for inadequate relief. With rare exceptions, relief effort by public and private groups tends to solidify the status quo without instigating basic changes in the society. Such efforts can, nevertheless, dull the pain of short-term suffering, prevent much human distress over the long run, and reduce the total toll of property loss by as much as 5–10 percent.

There is yet more to be learned about ways of acting quickly and in sensitive recognition of human needs, but the lessons enumerated above can be applied widely.

Insurance

Insurance may be a powerful tool for either good or evil, depending on whether two conditions exist in the policy offered. First, unless the protection is underwritten by solid financial resources beyond the immediate area of risk, an insurance plan against natural

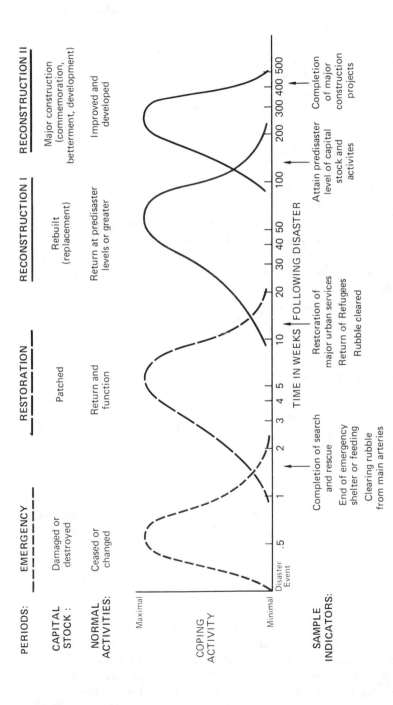

FIGURE 7.2. Relief and reconstruction after disaster. The phases of human response after an urban disaster are marked and distinctive but overlap considerably. Priorities of the emergency period are pressing and immediate as relating to elementary survival. In the restoration period they are opportunistic: the essentials of urban life that are repairable are restored as rapidly as possible. But the needs and opportunities of reconstruction are not clear and there is tension between the desire to reconstruct rapidly and the opportunity to build a safer and/or better city. This leads to two distinct periods: functional reconstruction to provide for the equivalent population of the destroyed city, and a commemorative or betterment reconstruction, which takes considerably longer. In general, each phase lasts about ten times the previous gap, and the institutions and modes of organizations appropriate for one phase may not be so for the others.

disaster will collapse in the wake of a catastrophe occurring early in its life or spreading across a large part of the nation. Plans for either re-insurance or public guarantees must be provided from the outset.

Second, a financially stable policy may still exacerbate the situation it is intended to remedy unless it is accompanied by premium rates proportionate to the hazard or linked with land-use regulations. Ideally, people who deliberately locate in the path of an avalanche or in an active seismic zone or within reach of a cyclone surge should be expected to bear the cost of the insurance as well as that of indemnification for loss. A government may, however, want to extend coverage — at a subsidized rate — to people already there. Such assistance is a dignified and possibly efficient substitute for some emergency relief and a means of alerting people to risk. In that event, insurance offerings must be guarded by strict requirements about land use to avoid effectively encouraging people to remain in or migrate to the risky areas. If not, the New Zealand record for promoting building on shifting foundations will be repeated on a grand scale (O'Riordan, 1971a). Properly designed and executed, insurance can spread the risk equitably and foster economic reduction in loss potential; improperly designed or executed, it can worsen the whole situation.

Preventing Effects

There is now enough experience with devices both technical and administrative for reducing the injurious effects of great natural disasters to allow us to draw a few sharp lessons. These relate to control works, building design, and warning systems.

Control Works

For most conventional control devices, such as sea walls, retaining walls, levees, storage and detention dams, and channel modifications, the basic problems of effective design and construction are well recognized. The need for planning a water-control works as part of a comprehensive basin program is established but not always implemented with skill. A substantial proportion of new control works fail to achieve their expected goals for one or more of these reasons: (1) the cost frequently exceeds the estimates by economically disastrous margins; (2) the need for planning auxiliary programs, such as land-use regulation in floodplains or rural credit in irrigation projects, is neglected; and (3) the full range of long-term environmental side effects, such as those on local ecosystems and on public health, is not investigated from the outset. Yet, there is no longer valid excuse for repeating these errors of the past. Guidelines are available to funding agencies. Professional advice can be obtained. Procedures for estimating environmental impact are coming into play. By all these means international experience can reinforce national experience.

The extreme view that technological fixes can curb all of the floods or hurricanes or landslides is discredited by tragic experience. Taken alone, such a fix is likely to encourage greater risk taking. Quite aside from technical feasibility, the cost would be prohibitive. In local situations, however, there are still ample opportunities for technical solutions that are cheap even when all social costs are figured in. They need to be weighed and refined and applied in proper balance with the other types of adjustment. In practical terms, perhaps a quarter of property loss could be saved by such measures.

Building Design

Engineering methods for reducing the vulnerability of structures to shaking earth, flowing water, and high wind are sufficiently advanced to warrant local or national specifications for buildings. Except for earthquake-resistant designs for high buildings, such methods have not been widely distributed or tested. Although beginnings have been made in earthquake engineering design, some features (such as the effects of earthquake on elevator operation and fire incidence in skyscrapers) require extensive research before they can be recommended indiscriminately. International review teams have started to operate under UNESCO auspices in areas torn by natural disaster, but the chief responsibility has been carried by engineers working through government agencies or as consultants who share experience through international professional societies.

No longer are there good reasons for unwittingly constructing in an earthquake or hurricane zone a building of a design known to have failed in an area of similar hazard, or for planting a crop that is susceptible to heavy loss when drought- or flood-resistant varieties are available. The information is available from one source or another. Ordinarily it is not disseminated and even if known to an elite few is slow to be applied by the builders or farmers who could use it. When we take into account the whole range of hazards, it seems likely that loss of life could be reduced by at least 10 percent and that property loss could be cut by about the same amount through changes in design of vulnerable property and land use.

Warning Systems

Nowhere on the globe in 1976 was there a warning system that incorporated all the technical and social devices known to be practicable and useful. The performance in issuing warnings continued to be uneven and everywhere incomplete. The opportunities for use of satellite imagery were not yet grasped, but even when they had been firmly in hand, the operating systems fell short, in a variety of ways, of applying fully what was known. Thus, forecasts failed to use effective observing equipment or computing facilities; dissemination of forecasts through media or persons was less than complete; where the forecast reached the right people at the right time, it was rarely accompanied by precise advice as to what to do. Most of such advice could be followed fruitfully only if there had been adequate preparation long in advance in the form of education, organization, and physical facilities (e.g., roads and sirens). In short, if warning systems as well as structural designs are to be established or improved on the basis of existing knowledge (assuming no breakthroughs in telecommunication and remote sensing capabilities), the global loss of property from natural disasters could be reduced by at least 10 percent. Loss of life could probably be cut to a quarter of the present toll registered for floods, cyclones, tornadoes, and tsunamis.

Modifying the Event

Although there have been waves of optimistic prediction that floods could be prevented or droughts corrected or hurricanes dissipated by technical measures, the performance so far, as illustrated by the Australian and the United States records, is modest indeed. Hence, it behooves the international community to exercise great caution in relying upon modification programs, especially those concerned with weather.

The opportunities for modifying flood peaks by upstream land-use measures are highly limited in most areas, and must be viewed as supplements to other adjustments. Again, the likelihood that breakthroughs in water desalting could change the character of arid lands is slim indeed. Nor is cloud seeding yet proven for hurricane modification; and while it holds some promise for hail prevention, it has only local significance in many drought situations. Moreover, the danger is now recognized that major changes in weather might generate much wider climatic and ecosystem disturbances, and should be approached only with great caution.

Land-use Changes and Relocation

While the historical importance of migration as a device to alter hazard interaction is well established, the opportunity to use it today is heavily constrained by political and ethnic barriers. Without drastic changes in working concepts of national sovereignty or of United Nations authority, the idea that farmers stranded in a drought-ridden sector of northern Mexico could pick up and move to irrigated oases in the United States is only whimsical. Notwithstanding its generosity in succoring fugitives from civil war in Bangladesh, the government of India has not considered a wholesale transfer of farmers from the hurricane-swept coast of the Bay of Bengal to its interior uplands. Great migrations of that sort are currently at an end, except in very dry areas of Africa, where nomadic herdsmen move their livestock with slight regard for national boundaries.

A different kind of migration sends out huge demographic waves — the movement of agricultural workers to centers of city employment. Though most common within nations, it sometimes involves movement across national borders. These streams of migrants seeking temporary jobs reflect the pressures of hazardous life in many instances, such as that of peasants traveling out of Brazil's northeastern "drought polygon" to distant cities. Learning how to stem these currents of humanity or to divert them from the city or to care for them where they come to rest remains a perplexing task in which nations can join.

On a local scale, the relocation of peoples within a climatic zone-from Sri Lanka floodplain to adjoining bluff, from Managua fault zone to more stable terrain — is largely catch-as-catch-can. Governments are only beginning to understand the social distress they generate when they relocate, say, a hundred thousand farmers out of a tropical reservoir site as in the Volta basin in Ghana (Scudder, 1968); and the notion of massive moves from risky to less risky zones appears utopian by comparison with immediate schemes for improving living conditions in place. One recent effort by the Tanzanian government to relocate farmers off the floodplain in the Rufiji River indicates to date that the moves along the main stream to higher ground were expedited by building new roads that led to provision of better market and social services. Although the new soil was somewhat less productive than what they had known, the gain in services appears to have offset the loss in soil fertility, and the people have remained on higher ground in contrast to those downstream (Sandberg, 1973, 1974).

In the Rufiji delta the people refused to move because they could see no viable alternative to cultivating productive, seasonally flooded land. Ways of enabling them to stay on the land without suffering serious loss remain to be explored. Basic study of the prevailing ecological system is necessary before an effective relocation can be achieved. Large-scale changes in location are at best a difficult enterprise. It may become necessary to carry them out in certain areas, but the complexities will severely tax the capacities of government. The tendency is to look for land management solutions in place.

Within urban areas the desirability of regulating or planning land use to avoid undue exposure to earthquake, flood, landslide, and hurricane is well recognized. That one sensible land-use plan is worth a dozen gallant rescue operations is beyond dispute; but putting the appropriate plans and regulations into effect is another matter. The first serious international exchange on floodplain regulation took place in the Soviet Union in 1969; the first international discussion of microzoning of earthquake areas was convened in the United States in 1973. The ideas are gaining currency but are far from adoption in all but a few countries.

It is difficult to calculate the consequences of thoughtful, long-term land-use planning and regulation for the world's vulnerability to hazards. At worst, such a program could reduce productivity if applied blindly or incorrectly. As a minimum advantage, it could halt the rising toll of loss. As a maximum, it could progressively restrict the occupation of hazardous zones to regions where economic gain clearly exceeds economic cost to the nations affected, and thus eliminates damage to the economy.

Study Methods

In view of the immense diversity of hazard conditions, it would be impracticable for scientists to prescribe suitable methods to apply in determining what mix of adjustments would best suit all local conditions: many forms and styles of investigation may be in order, and the range in cost is formidable. Nevertheless, two general principles seem to emerge from the multitude of studies completed to date. One is that it always is desirable to indicate, if only in the sketchiest way, the nature of the interlocking systems that produce the hazard. Rarely is it possible to identify, let alone accurately measure, all the components that would be affected in managing a coastal area or in altering the pattern of airborne toxic substances in a city. For example, great scientific skill and data stores were required for the quantitative and comprehensive systems model of the relation of weather, forest stand, budworm pest, timber production, and recreational land use developed by Holling, Jones, and Clark (1976) for spruce forest lands. Nevertheless, rudimentary efforts continue to be made.

Second, the record of ambitious projects that failed to take account of the aspirations, skills, and needs of the people affected shows that any study has to give some attention to existing adaptations and adjustments as a groundwork for suggesting changes. It is not enough to posit the fact that people suffer from drought or cyclone or flood, and then prescribe remedies. A basic ingredient is understanding of why people carry out their current practices and how they perceive their environment and the options open to them.

These are the first principles that could be shared on a wide front, with a corresponding reduction in the impact of extreme natural events upon society. Applied, they could lead to substantial reduction in both loss of life and in property damage (Figure 7.1). To put them to work requires initiative on the part of public agencies, either through direct action or through information and education. Much of the work, as shown in Chapter 6, can be undertaken by nations acting independently.

WHAT IS WORTH DOING JOINTLY?

Although, strictly speaking, all the fruits of natural-hazard experience can be shared in some fashion by way of international collaboration, certain types of action are not likely to flourish unless encouraged by cooperation across national boundaries. Seven steps are worth taking jointly within the family of nations. Three of them are already partially developed: (1) there is now a loose network for monitoring extreme events; (2) a framework for coordinating disaster relief is established; and (3) a beginning has been made in exchanging research findings on the processes and constraints involved in natural hazards.

The other four channels for collaboration are in the early stages of discussion: There is more hope than performance in (1) developing international warning systems, (2) assessing social losses, (3) providing technical guidance in hazard management, and (4) encouraging working schemes for insurance coverage. All these measures are based on scientific research that is scattered and largely uncoordinated. Research undergirds a few activities (such as warnings), and additional research will be stimulated by concerted efforts to act along the other lines.

Global Monitoring

National governments have long maintained scientific services to observe and record extreme natural events within their own territory. Monitoring on a global basis, however, is more recent, and was motivated by three needs: (1) the need to extend time-limited observation records for natural phenomenon made within a country by studying those observed elsewhere; (2) the need for large quantities of comparable data for forecasting and research; and (3) the need to monitor the newly recognized environmental threats that are global in origin, transport, or effect (National Academy of Sciences, 1976).

Numerous descriptions of great droughts, floods, storms, and earthquakes are found in historical accounts and records, but accurate physical measurements of such extreme natural events are hard to come by. Few streams in the world have precise, valid measurements of streamflow longer than that of the Colorado River in Mexico and the United States, where the U.S. stream-gauging program was established in 1894. Many longer records exist but are inaccurate for one reason or another. For example, Nile River records go back to A.D. 622 but were politically manipulated, since tax rates were based upon the river heights reported (Hurst, 1952). Systematic rainfall measurements are available only since 1860 in the United Kingdom, in which nation they were first taken — although more primitive measurements began as early as 1677. Earthquake recording by seismograph dates back to 1897. By piecing together historical accounts, we can estimate the magnitude of past earth tremors. In the People's Republic of China this search reaches back two thousand years.

In cases where scientific records are not in hand — and this includes a large proportion of the land surface and an even larger part of the seas — it may be possible to infer earlier phenomena from physical or biological evidence. After a great downpour in which the maximum precipitation recorded in a rain gauge was 25 centimeters, hydrologists scour the countryside for buckets, boats, or other containers revealing that depths of 20–35 centimeters occurred at the same time within a radius of 30 kilometers. Driftwood lines on the shore delimit an earlier flood; collapsed walls reveal unobserved wind velocities; the size of boulders indicates the force of a torrent; rings in trees suffering from lack of moisture record the severity of a drought. Scores of other rough indicators may be used to estimate the reach of a tropical cyclone, the shaking duration of an earthquake, the velocity of a flood flow, the severity of a drought. Such inferences may grade in time horizon from estimation of yesterday's tornado, when meteorologists examine the size of a log on a roof, to reconstruction of a dry spell a thousand years ago by measuring a clay varve or the rings of a gnarled pine tree.

Given the problem of inference, the initial response of some scientists to natural hazard is to step up their appeals for programs that will provide more and better data. For some hazards this may be very much the need as shown in Figure 7.3a for earthquake strong-motion recording (Page and Joyner, 1974), or quite the opposite, as in the case of small stream gauging (Carrigan and Golden, 1975) shown in Figure 7.3b.

In practice, most calculations of the probability of an event of extreme magnitude use whatever inferential historical data can be collected to estimate possible magnitudes, and to analyze the recorded measurements to set probabilities or recurrence intervals for those events. The task calls for careful and prudent thinking about chance processes. From 30 years of record, a hydrologist attempts to assign probabilities to events ranging up to 0.2 percent (1:500 years) or 0.01 percent (1:1,000 years). Synthetic sequences can be generated through use of randomized computer techniques, and refined statistical analysis will help; but basic evidence is slim. In these circumstances the broader the areal coverage of random events, the more the record is strengthened to offset the short time span available for analysis. By aggregating short-term data for several similar areas, the depth of coverage is in effect extended.

The major impetus to monitoring on global scale has paralleled world concern with the impact of human activity on the environment. The most effective monitoring effort is still the quasisecret Great Power network to track nuclear explosion and radiation products. More recent has been a concern with carbon dioxide (CO_2), pesticides, particulate matter in the atmosphere, heavy metals, and a variety of man-made pollutants. Then, too, the Stockholm Conference on the Human Environment placed heavy stress on keeping track of changes in the quality of environmental components — air, biota, soil, and water — with interest centered on whether significant deterioration is in progress. In the case of

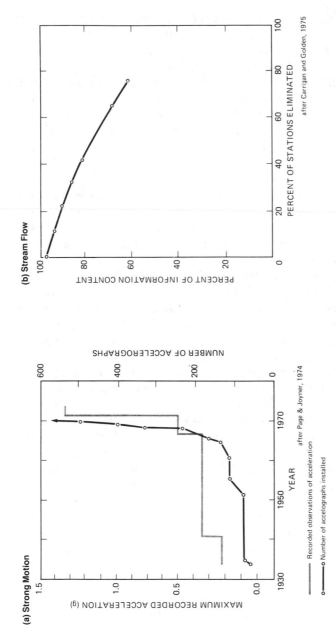

FIGURE 7.3. Data: Too little, too much. From 1933 to 1966 the peak horizontal acceleration recorded in the U.S. was 0.36g—substantially below the value of 0.5g commonly accepted by designers of buildings to represent the most severe ground motion that would occur in even the largest earthquake. As more strong-motion instruments were placed closer to active faults, much larger accelerations have been recorded (Page and Joyner, 1974). In contrast, better analysis of streamflow data can permit major reductions in the size of a network with only limited loss of information. For a small stream network in Illinois, the 47 gauges could be reduced by half and still retain 80% of the information provided by the entire network at a savings of perhaps $400–$1,200 per year per station (Carrigan and Golden, 1975).

200

natural hazards, attention may be expected to turn to three questions:

1. What is the magnitude and probability of the unusual event?
2. Over recent centuries or decades, are there any trends in these characteristics that can be identified?
3. Are there any premonitory or predictive indicators for the onset of disaster?

Answering these questions calls for a good deal of accurate, persistent observation and analysis. These, in turn, require international networks and organizations capable of sustaining both data collection and inquiring minds. Routine recording unaccompanied by imaginative study can fail to grasp what is happening (see Figure 7.4).

The most comprehensive network for observation of related physical phenomena is the World Weather Watch of the World

INDICATOR OF CLIMATIC CHANGE:

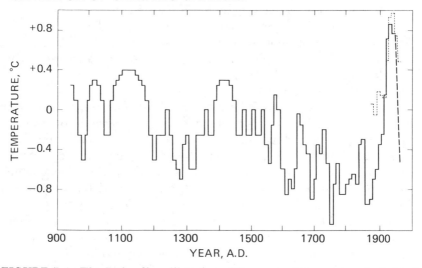

FIGURE 7.4. The Icelandic millennium. Mean annual temperature in Iceland over the past thousand years, as reconstructed by Bergthorsson. The heavy dashed line indicates the temperature change of the last decade, and the dotted line shows the change for the Northern Hemisphere in the last century to the same scale. The climatic "norm" currently used is the 30-year average of 1931–1960. This "norm" seems to be the most abnormal of record, and if the Icelandic record is an indicator, perceptions of climate and weather derived from this period may prove very misleading (Bryson, 1974).

Meteorological Organization (WMO). Incorporating surface climatological measurements from 25,381 stations in the 126 member countries with upper-air and other observations, the data coordinated by WMO provide information on temperature, precipitation, wind velocity, barometric pressure, and related parameters of tropical cyclones, tornadoes, windstorms, floods, heat waves, cold waves, snowfall, avalanches, and other events related to weather.

Less comprehensive data networks exist for other phenomena. For streamflow, collection of data in comparable form was strongly encouraged by the International Hydrological Program sponsored by UNESCO. The initial decade spurred observations in some places, and fostered cooperation among hydrologists with related problems. In the mid-1970s in a basic "decade" network there were an estimated 1,500 stream-gauging stations, 260 lake stations, 640 pan-evaporation stations, and 320 lysimeters to measure soil moisture. During the 1965–1975 decade the number of observation stations more than doubled in a number of countries, among them Australia, Ghana, and Nicaragua. The need for river data transmission is far less urgent than that for weather data, except where a stream crosses national boundaries.

A center for collecting data on tsunamis is maintained in Hawaii by the U.S. Department of Commerce. The International Association of Volcanology and Chemistry of the Earth's Interior, located in Rome, maintains a Catalogue of Active Volcanoes, and has described 619 volcanoes as to location, form and structure, activity, and petrography.

In the United States the Earthquake Information Center in Boulder, Colorado, records readings from seismographs operated in this nation as well as in other nations. The records on earthquakes in Europe and the Mediterranean Basin are maintained at Strasbourg, France, by the Bureau Central International de Seismologie. In 1974 the Earthquake Center made preliminary determinations of 5,007 epicenters. Data on storm surges are collected by the Institute of Coastal Oceanography and Tides at Liverpool, England. UNESCO publishes annual summaries of information on earthquakes, tsunamis, storm surges, and volcanic activity. Out of this complex and somewhat ponderous array of agencies dealing with geophysical hazards, the scientists of member countries can draw information giving perspective on their local problems.

With the development of meteorological satellites the transmission of weather data is speeded, and with the Earth Resources Technology Satellite the opportunities for recording and processing of these and other geophysical and biological data are expanded.

Yet, up to 1976 an international system was still in the making. Such a system could replace some of the activity of individual agencies; it has, moreover, the potential of disseminating certain observations widely and cheaply.

Under the sponsorship of the Smithsonian Institution, a Center for Short-Lived Phenomena was established in 1968 to identify and report on ephemeral phenomena of geophysical, biological, and anthropological character. These include the convergence of 3 million blackbirds on Scotland Neck, North Carolina, in 1970, and the Victoria, Australia, mouse plague of 1969, as well as the more conventional typhoons and earthquakes. More than 3,000 correspondents in 144 countries are linked to the center by a telecommunications network inherited in part from the satellite-tracking program mentioned above, using the International Radio-Relay League's amateur network. In a recent year it reported about 150 unusual events, one-third of them classified as natural disasters.

There is probably no scientific need to encourage international management of a comprehensive monitoring system restricted to physical features of natural hazards, since it can be expected that as basic observations of physical phenomena are expanded, understanding of extreme events will thereby increase. The case for comprehensive schemes is particularly weak in many developing countries where the observation networks are thin and short-term. In time, if national experience is any guide, monitoring efforts will need to be extended by enlarging national capacity to make observations. Meanwhile, there is more urgent need for intergovernmental agencies to provide professional services to nations in which the development of networks and analytical skill will be a slow process. The aim of such services would be to aid scientists and technicians who are limited to estimates as a basis for designing adjustments and lacking time, money, and skill to carry out sophisticated analysis of experience outside their own areas of study.

Forecasting and Warning Systems

When observed data are used to forecast extreme events and to warn the people affected, the results may be immediate and drastic. The distinction made in Chapter 2 between issuance of forecasts and the operation of a complete warning system is important. An entirely accurate and timely forecast of an extreme event may have little or no effect upon the event's victims if the warning system is inadequate. Until the original forecast is transmitted to the people affected in an intelligible form, at an appropriate time, and with ample preparation to enable them to act constructively, it is not a genuine warning. Many, if not most, of the publicized warning schemes fail to provide essential links in this chain between scientific prediction and human response. Their directors often remain comfortable in the knowledge that one link in the system — their link — worked well, although the system as a whole failed to achieve its goal of reducing human distress.

Ordinarily the actual forecasts as well as the related components of a warning system are the responsibility of national agencies. Thus, it was the Pakistani Meteorological Service that prepared the November 1970 cyclone forecast. To do so, however, required international cooperation in the sharing of observations and of forecasts from neighboring countries. In that instance India, Burma, and Thailand, working with technical assistance from the WMO, were involved in putting together the picture of the great disturbance moving up the Bay of Bengal. Again, Japan maintains a thoroughly competent facility for forecasting tropical cyclones in the western Pacific, and it cooperates in similar efforts in the southwestern Pacific. Wherever the phenomena cut across national boundaries, there is opportunity to strengthen the forecasting ability of each nation by pooling the information from all. A number of other promising beginnings are under way in developing genuine warning systems at the international level. The Caribbean area has an active, comprehensive system that accurately identifies and charts virtually every tropical cyclone in the region. Without benefit of treaty, the various meteorological services exchange their observations through the Hurricane Research Center in Miami, supported by the U.S. National Oceanic and Atmospheric Administration.

Almost routine is collaboration in forecasting river floods for the purpose of improving the management of river works and of evacuating flooded areas. Except insofar as rivers cross national frontiers, there is little need to provide international warning services. National forecasting agencies are linked under WMO auspices to exchange meteorological observations upon which flood forecasts are based. In Europe there are long-standing agreements covered by treaty for exchange of streamflow and stream elevation data among countries in the Danube and Rhine basins. The United States and Mexico collaborate on the Rio Grande, as do the United States and Canada on the Columbia River and the Great Lakes. With the scientific assistance of UNESCO, a cooperative flood-forecasting system was developed in the late 1960s for the Lower Mekong River by Cambodia, Laos, Thailand, and Vietnam. This system is possible only because of its joint construction and joint maintenance, under the Coordinating Committee for the Lower Mekong, of a hydrologic observation network. The network survived the encroachment of ground warfare and bombing with remarkable success during the period 1965–1973, and serves both in the design of new projects and in forecasting for river management.

Yet there are scores of international rivers where forecasting is not a joint effort, and hence the opportunities for formal collaboration in identifying droughts are large. The cooperation so far generated has occurred for the most part where there was already a framework for cooperation, where the need was unambiguous, where technical proficiency was assured, and where the work did not require elaborate commitments, such as on the Columbia and the Rhine. The precedents are now established, and similar arrangements may be expected elsewhere.

The International Tsunami Warning System was devised in 1948 to furnish forecasts for Hawaii's shores, but it has extended its scope to virtually all the Pacific Basin nations. The very instant that a seismic disturbance severe enough to produce a tsunami is measured in the Pacific Basin, it is observed at or reported to the Honolulu Observatory. The epicenter location is determined and, if a tsunami seems possible, a tsunami watch is issued. Time of arrival on affected shores can be predicted, and this prediction tested at once in scattered stations against observations of sea level. Warnings, whether local or general, are passed on to national agencies.

The Honolulu forecasts are supplemented by those of national and regional agencies such as the Chilean Navy and the Japanese Meteorological Agency, which take responsibility for issuing more detailed warnings. Japan alone has five tsunami forecast centers that function with virtually full responsibility for spotting tsunamis generated in nearby seas. To the extent that it becomes practicable to predict earthquake by measures of dilatancy or other precursory phenomena, as claimed in the People's Republic of China, more elaborate networks for recording seismic phenomena will be required; and, as these techniques improve, forecast systems may be expected to expand and develop.

A forecasting system is an essential element in the international program for control of desert locust invasion in the arid and semi-arid region between Afghanistan and Morocco. With political support and stimulation from the U.N. Development Program (UNDP) Special Fund in 1959, and with the collaboration of the U.N. Food and Agriculture Organization (FAO) and WMO, forty-two nations shared in a program of survey, reporting, forecasting, and control that seems to have largely eliminated the pest (Townley, 1968). The meteorological warnings told where and when to expect movement of the swarms. The system worked so effectively that the locust movements were controlled over large areas, and the need for the system thereby diminished.

An attractive possibility sometimes mentioned in the same breath as new satellite communications and monitoring systems is that of using earth resources and surveillance satellites to forecast disaster situations. At present the dream exceeds performance, and there is doubt as to precisely how satellite imagery and reporting can drastically enhance the capabilities of the warning systems already in operation. Thus, when a tropical cyclone is located in the Gulf of Thailand or a severe storm in the lower Parana River, the critical question is not whether it exists or how long it endures: those facts can be gleaned from ground reports. The crucial problems concern internal wind velocities or rates of precipitation, and their solution requires airborne, radar, or surface measurement. By 1976 the hope expressed at Stockholm for an improved international warning system based on satellite observations had not materialized, but the possibilities were being explored more carefully.

Assessment of Losses

Without valid ground for estimating the social impact of hazard, a judgment as to what measures are worth taking may be highly skewed. One sound basis for calculated action to deal with a natural hazard is accurate knowledge of the magnitude of property damage and loss of life; but perhaps even more pertinent for the evaluation of policy is knowledge of trends and of the rate of change in such losses. Except for the instances cited and a few other situations, the magnitude of the loss and the trends cannot be reliably estimated from present data. As a case in point, our own preliminary estimates of the global incidence of loss from natural disasters that appear in Chapter 1 are gross approximations and give little detail. An annual summary was initiated by UNESCO in 1970, but this was undertaken on a very moderate scale and is too recent to yield trend data. The effort has been terminated by its top administration, and UNESCO has not taken leadership in the field. It has, however, conducted a series of useful case studies of earthquake disaster, centering chiefly on the relation of loss to engineering structure. A new mandate laid upon the U.N. Disaster Relief Coordinator to expand, however modestly, the assembling of comparative statistical data on a global basis would be highly productive. These data might be incorporated in the U.N. Statistical Yearbook.

The pressing need is not for direct collection of field data by any U.N. agency, but for criteria for evaluating losses and methods of conducting surveys so that the results can be shared and compared among nations, especially for the benefit of developing countries, with their relatively short experience. Accuracy of aggregate figures of world loss is not of prime significance; it would be more persuasive to eliminate any doubt about the savings that might occur because of a cyclone warning system in a particular region, or about the categories of building accounting for greatest loss from an earthquake in a specified city. Even in high-income countries these data are for the most part lacking, though a few countries, such as Japan, have relatively detailed coverage of buildings affected by hazards, and where that coverage is lacking, the services of engineering and survey groups can be enlisted to provide estimates. Low-income countries have neither the data nor the technical personnel to collect it.

Comprehensive reporting on all losses experienced from a particular event would be useful, though hardly essential. Indeed, elaborate collection on a world scale would likely clutter up the process of discovering the type and magnitude of the effect generated by an extreme event. A thin and spotty national census of hazard data would be far less useful than careful inquiry into the hazard itself in selected places so that causes of loss and modes of dealing with loss can be recognized accurately. In high-income countries the indirect effects of hazard — the secondary or tertiary consequences for human organization and well-being of direct property damage or personal injury — are rarely traced; and yet a few searching investigations among such nations would enhance the estimates in developing nations as well. The fact of variation from place to place in culture and habitat would make it desirable to conduct studies of the type described above in representative situations, as has been explicitly encouraged in the Man and the Biosphere Program (UNESCO, 1973). To the extent that these are parallel among countries — as with the vulnerability of industrial plants to tornado or the gains to peasant agriculture from flood warnings — there are benefits to be shared.

Such assessments would be most fruitful if they were made both immediately after an extreme event and then again a year or two later. It takes time for the cumulative gains and losses to show themselves.

Studies of Process and Constraint

Scientists have been quicker than their governments to collaborate across national boundaries in dealing with hazard. For the most part, the tsunami, hydrological, and seismic information programs began with nongovernment scientists and then gained government support. Meteorologists, as an almost fraternal group transcending political boundaries in search of improved grounds for forecasting, have been the leaders in global collaboration.

To the extent that the WMO already speaks for the meteorologist and that members of the International Council of Scientific Unions (ICSU) already speak for the larger professional groups, there is lively cooperation in the mutual deepening of understanding of the basic physical and biological processes involved in hazard. The most ambitious and genuinely international of environmental research efforts is the Global Atmospheric Research Program (sponsored jointly by WMO and ICSU), which aims at exploring the major patterns of global atmospheric circulation. From this should come greater facility in anticipating the occurrence of extremes and of regional shifts in weather phenomena.

The International Hydrological Program (IHP) stimulated governments and private investigators to advance and collate observations on surface water flow, with accompanying gains to the capacity to estimate flood magnitude and frequency. The International Biological Program and its governmental successor, the Man and the Biosphere Programs coordinated by UNESCO, provide another effort to extend our knowledge of biological process to critical human environments.

Until recently, determination of the likely social pay-off from additional research on extreme events was largely contingent upon which professional group or scientific agency argued most persuasively that its line of work deserved to expand. Typically, each unit cited the toll of damage and lost lives as justification for its program. In informational programs, for example, it was seldom stated — as it is now — how the improved information would necessarily reduce uncertainty to the degree justifying its cost, and whether a reduction in uncertainty would lead to social behavior that will reduce the toll of lives and property lost. The possible social gains from research on natural hazards can be no greater than the possible prevention of losses plus the increase in productivity resulting if the new knowledge were put into effect. Thus, any judgment as to how much research it is desirable for a nation to support would depend upon its cost and its practicability in relation to those possible benefits.

Estimates of this sort require basic assumptions about natural processes and technology, but hinge inevitably upon the capacity of social systems to apply research results. To date the most detailed analysis of this question is the Assessment of Research on Natural Hazards in the United States, carried out under the auspices of the U.S. National Science Foundation. This work appraised for all geophysical hazards the likely effects of new inputs from research in the natural and social sciences (White and Haas, 1975; see Table 7.1).

On the international scene, the value of such assessments and of research inspired by them lies chiefly in the opportunity they provide for nations to learn from the mistakes of others and to pool their analysis of the significance of events of very low frequency. Thus far there has been no great effort to investigate which lines of research or action on natural hazards would offer most for developing countries. Some lessons from the U.S. assessment will be directly transferable to developing countries. For example, it attested that improved means of designing buildings against wind stress would have wide application. Other findings would have no relevance, and the reasons they are irrelevant — such as differences in methods of land-use control — need to be recognized by those who seek to transfer U.S. experience to other countries.

Disaster Relief and Reconstruction

Cooperation in disaster relief — typically, in the form of one nation's sending aid to another nation in distress — is at once the most dramatic, obvious, and highly developed aspect of the present pattern of international response. For reasons that are not entirely clear, aid to an earthquake victim is unusually enthusiastic and initially nonpolitical. The headline "10,000 Lives Lost in Quake" is followed within minutes by announcements that food and medical supplies are on the way and that a public collection is being taken for the victims.

Until the 1920s, disaster relief was delivered on a strictly bilateral basis — from one nation to another. Then, around 1922, the International League of Red Cross and Red Crescent Societies began to play a coordinating, intermediary role. During World War

TABLE 7.1. Research Opportunity Payoff Tableau

Research opportunity set	Net direct cost benefit-cost ratio ($)		Reduction of fatalities	
	Average value	Catastrophe value	Average value	Catastrophe value
Hurricane modification				
Hurricane dynamics				Low
Technology	>1	<1	High	or
Socioeconomic effects				negative
Insurance				
Formulation				
Hazard mapping methods	>1	>1	Medium	Medium
Adoption				
Land–use management				
Hazard mapping methods				
Adoption				
Socioeconomic effects	>1	>1	High	High
Hurricane proofing				
Technology				
Adoption				
Relief and rehabilitation				
Socioeconomic effects	1	1	Medium	Medium
Formulation				
Warning systems				
Hurricane dynamics				
Hazard mapping methods				
Evacuation methods	>1	<1	High	High
Social response				

II, semipublic citizen groups came into new prominence in providing for war sufferers, though small-scale efforts such as those of the Quakers were overshadowed by British War Relief in the United States and Oxfam in the United Kingdom. Many religious bodies set up their own relief agencies, and the Caritas and World Church Service provided a global framework for, respectively, Catholic and Protestant efforts that had begun in response to wartime needs and then extended to natural disasters.

On the heels of World War II, governments moved into a larger relief role than before. Bilateral aid programs, aimed at either reconstruction or economic development, included substantial allotments for emergency relief. The deeper involvement of governments increased the likelihood that they would seek to work with each other through an intergovernmental agency. Accordingly, the U.N. Children's Fund was created in 1946 to aid in food, housing,

Reduction of social disruption		Protection/ enhancement of natural environment	Contribution to equity of income distribution	Expected success of research	Political feasibility of adoption
Average value	Catastrophe value	Average value	Average value		
High	Low or negative	Medium	n.a.	Low	Medium
Medium	Low	Medium	Medium	Medium	Medium
High	High	High	n.a.	High	Low
Medium	Low	n.a.	High	High	Low
Medium	Medium	n.a.	n.a.	Medium	Medium

A format for making decisions as to research opportunities is shown in the Payoff Tableau. Sets of opportunities for research on tropical cyclone adjustments in the United States are analyzed comparatively for their potential utility, in terms of reducing damage loss, fatalities, and social disruption; enhancing the environment; and distributing benefits equitably. These are then compared vis-à-vis the likelihood of research success and the possibility of adoption of the adjustments.

medical, sanitation, and other services for the well-being of children only; some 4–5 percent of its expenditures go for emergency aid. Again, the World Food Programme, set up by FAO in 1963, distributed more than $800 billion worth of food during the next nine years, between 30 and 85 percent going to victims of natural disasters.

Equipped with noble instructions and a budget for only a handful of staff, the U.N. Disaster Relief Coordinator set up his base in Geneva in 1972 and began to find out what activities might benefit from coordination. The first major test of the capacity of

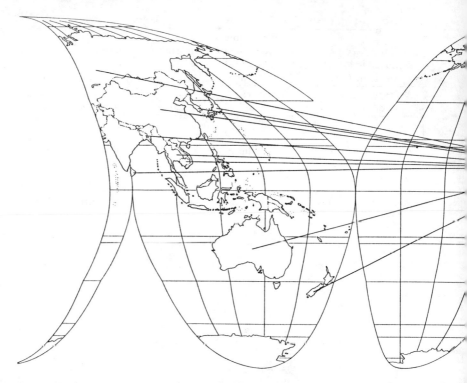

FIGURE 7.5. Global response to disaster: Origin of aid to Managua, Nicaragua. The response to a major disaster is truly global. Six months after the earthquake of 23 December 1972, in Managua, Nicaragua, at least fifty-one nations had sent aid through governmental channels and twenty-three more through the Red Cross or other non-governmental agencies. All in all, over $65 million in cash or services was contributed and perhaps three times as much pledged in loans and credits toward reconstruction exceeding $400 million.

UNDRO was the Managua earthquake of December 1972. Large amounts of bilateral aid were provided, and coordination was urgently needed. An excess of temporary housing was promptly built in Managua, and a year and a half after the earthquake only 6,500 families had moved into units prepared for 10,600. In the long run, however, the project may prove a boon for the poor of Managua, as plans to convert the temporaries to permanent low-cost houses rapidly take shape (see Figure 7.5).

A second area of concern — and one about which little is known — is the possible diversion of development aid to disaster relief. During the Sahelian drought in sub-Saharan Africa, U.S. aid

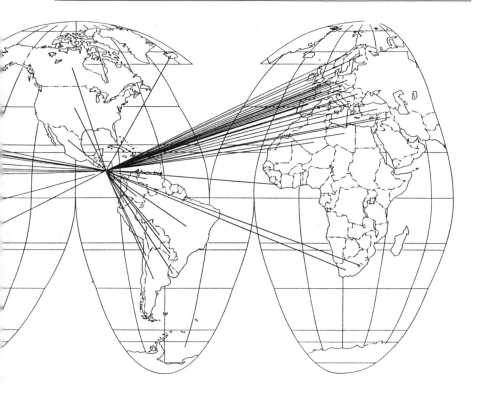

was limited largely to relief, to the exclusion of long-term development there and elsewhere for which funds were formerly available.

A third area of concern was brought to prominence during 1975 by apparent government suppression of information about drought-induced famine in Ethiopia, although similar issues had been raised relative to disaster in the eastern region of Nigeria and elsewhere. Episodes such as these lead observers to inquire what the responsibilities and mechanisms of the international community for disaster relief are when a national government ignores it and does not request assistance.

The pattern of international involvement in relief and rehabilitation unfolds along similar lines in most countries that have been studied. Voluntary effort begins with citizen groups at the national level and moves toward international collaboration. Government effort also begins at the national level, and bilateral activities are supplemented by specialized multilateral agencies such as FAO. These, in turn, are coordinated through UNDRO.

At the national level, as demonstrated in the United States and Australia, assumption of disaster relief by a higher order of government was the prelude to later, greater involvement in activities other than immediate relief, as in the case of preparedness planning in the United States. Similarly, the establishment of UNDRO is probably the precursor of a much expanded program of international activity, though it may concentrate initially on organizing such relief. Nevertheless, it is clear from the outset that a program along the lines noted above, valuable as it undoubtedly is, is by itself insufficient to cope with the toll of losses.

Hazard Management and Disaster Prevention

The present tendency at the international level is to emphasize cooperation for scientific understanding and forecasting of physical phenomena while seeking to coordinate this with disaster relief. These two emphases can lead to a situation in which the peoples of the world gain a better understanding of the environmental processes that produce disasters and become more effective in providing relief when the dreaded event occurs. Some modest reduction in damage might also be achieved by improved forecasting and warning systems. But the outstanding need is to avert the catastrophe.

In the modernizing sectors of traditional economies, undoubtedly the highest priority in hazard management is to prevent or reduce the potential for further damage. It is there in the floodplains of Sri Lanka, the new croplands of Tanzania, and the expanding borders of Managua that vulnerability to extreme events is increasing most rapidly. There, too, the possibly disruptive effects upon world political economy may be most severe. Immediate activities should seek at a minimum to curb expansion of unduly risky situations. This requires the cultivation of new or improved understandings among the personnel of both international agencies and national governments.

With rare exceptions, administrators and technicians have been trained to cope with disaster rather than to prevent it. They understandably tend to concentrate upon the skills that they have developed, and their established procedures support that stance. Typically, the designer of a tropical cyclone forecasting program does not expect that the program will involve planning for altering land use, building design, or control works along the same coast. Again, specification of standards for emission from the stacks of a manufacturing plant rarely canvasses the relative social costs and benefits of alternative ways of meeting the needs satisfied by this factory. The study instructions reflect this view.

Guidelines

Instructions for preparation and review of proposals for transnational funding of development projects are one means of shaping initiatives in disaster prevention. Every bilateral and multilateral agency has its established procedures. These instructions can be so written as to ensure at least minimal consideration of two questions. The first is whether such and such a proposal for expenditure has involved the critical examination of the whole range of alternative measures for dealing with the hazard. As already outlined, these would cover such conventional measures as engineering control works, forecasting and warning systems, resistant building design, land-use management, and insurance.

The second question is whether estimates have been made as to the changes likely to take place in the environment and in hazard vulnerability as a result of forces still at work in the area. There is little point in investing in flood protection in one reach of a river if the human invasion of floodplains downstream promises to continue at an active pace. Well-intentioned aid in rebuilding a city damaged by earthquake is of doubtful value if reconstruction leads to even greater damage in the next massive earth movement by concentrating new buildings in risky areas. Many such changes are difficult to predict, but an attempt can be made to identify and project the more obvious trends.

National and international agencies have both had enough experience preparing environmental impact statements that guidelines in these directions can readily be drafted and gradually applied. Such guidelines can hasten the time when responsible administrators get to the details of altering their development plans in time to avert potential catastrophe. High-income countries have begun to prepare such instructions (see Chapter 5). Ironically, in the meantime they are busy exporting antiquated approaches that will worsen the situation in areas that can least afford the burdens of loss and adjustment costs.

Training

A series of international workshops and training sessions would be necessary to encourage national policy makers and administrators to develop national hazard policies and to make use of the new opportunities provided by United Nations initiatives. These might be difficult and unproductive in the beginning, since, as we have observed before, most administrators and scientists are trained in Western universities and government agencies steeped in the tradition of the technological fix, and hence could feel uncomfortable, and ill-prepared to try other alternatives, as demonstrated by relief administrators in Tanzania or flood administrators in Sri Lanka, who apply standard solutions to local problems. The unimaginative executive tends to do what he knows how to do, even though it may turn out to be counterproductive in time.

When the United Nations and the then Soviet Union brought together responsible technicians from developing countries at Titis in 1969 to consider alternative ways of reducing flood losses, the initial response of the participants was defensive: exclusive reliance on levees, channels, and dams was not to be challenged. In time, however, new views were considered and adopted by national agencies. But, of course, incentives beyond fellowship and education and short-term training are required to support shifts in administrative approach.

Development Plans and Proposals

A prime motivation for such shift might come from the policies of multilateral and bilateral assistance agencies. Procedures could be set up by funding agencies for a systematic review of a nation's development plans and proposals so as to recognize the moment when increased liability to loss from natural hazard threatens the course of economic development. Review along this line could be part of appraisal of a nation's plans formulated in the UNDP and by the International Bank for Reconstruction and Development (IBRD). Bilateral aid programs could develop similar modifications in their review procedure for national projects.

Steps toward adoption of the above proposals should be undertaken immediately, since they are likely to be very slow in materializing. Professional training runs counter to such proposals, and any modification in already unwieldy review procedures is unwelcome. It is difficult enough to think of the commonplace of the future without calling up the extraordinary event.

Insurance

There is no international insurance against loss from natural hazard. Lloyd's of London is the chief exchange for the purchase and trade of special policies, yet Lloyd's volume of hazard coverage is minute by comparison with world losses through extreme events. As the potentially positive role of insurance in guiding hazard management becomes more widely known, national interest in insurance will grow. Smaller countries then will face the extremely difficult task of setting up a plan that would not be annihilated by a great catastrophe during the early years of protection. In that event the possibility of some kind of re-insurance on an international scale will arise as a substitute for multinational relief and reconstruction. The sooner the practicability of such ventures can be investigated, the wider the set of options open to developing countries. The options should include multinational organization of insurance offerings as well as multinational cooperation.

Global Accident and Conjunction

Unless carefully managed, an integrated network of organizations to prevent natural disaster or to succor its victims in vulnerable regions or nations might reduce localized suffering while sowing the seeds of world catastrophe. Just as effective relief unaccompanied by other measures can encourage larger loss in the flood or cyclone of the century, major disasters could be cultivated on a world scale. To the extent that smaller nations become dependent upon larger nations for help and that communications and flow of goods become complex, with small stocks of stored food permitted by sophisticated planning, the possibility of severe, widespread distress may be enhanced. Probability of occurrence would be reduced while the magnitude of a possible collapse increased. Were all countries to learn to draw upon the common stocks when disturbances are triggered by extreme events, they might cover ordinary fluctuations at low cost until a highly unlikely combination of troubles in several areas threw the whole arrangement out of gear.

Drought spreading over a large territory, floods destroying stored crops, tropical cyclones crippling transport facilities—any one of them complicated by strikes, revolution, or wars—could bring tremendous disarray in social programs. The worst famine of recent decades—that of Bengal in 1943, during World War II—cost the lives of about 2 million people, primarily because ocean shipping was diverted by military strategists from rice importation at the same time that native crops were failing.

As we have noted earlier, three kinds of unforeseen events could trigger catastrophes that are worldwide in effect, in contrast to recorded local and regional hazards in nature that do not ride the

Apocalyptic horses of war, famine, or death. Thus, (1) a long-term shift in climatic belts could provoke a series of droughts or floods of historically unprecedented magnitude reaching around the globe. Or (2) a random conjunction of an extreme event in one area with a different extreme event in another—for example, a Chilean earthquake with Chinese drought and Ganges flood—could overstrain the capacity of the world economic system to respond. Or (3) a failure of a technological remedy—for example, the unexpected susceptibility of a new cereal variety to a virulent plant disease—could render extensive areas vulnerable to destruction from otherwise moderate floods and drought. To the extent that the measures indicated for dealing with the usual course of events are adopted, the impact of the global catastrophe will be reduced. More likely, however, those measures will not be adopted fully until catastrophe etches their need in suffering.

CHAPTER EIGHT

Natural Extremes
and Social Resilience

From Char Jabbar to Wilkes-Barre, from the Usambara Hills to the south Australian plain, the world is a mosaic of situations in which human settlement creates both resources and hazards. When we look in detail at Mexico City or Manchester, it is the uniqueness of the interaction of natural event and human use in each place that dominates our understanding of the response to extreme events. When we step back, however, it is possible to discern a few common elements. Our observations of eighteen countries and scattered places within them point toward a theory of what happens when people generate hazard, and the need for theoretical understanding such as this is one of practical urgency. A bewildering array of environmental threats having catastrophic potential appears on the world scene with increasing frequency. As thoughtful people everywhere strive to comprehend the complex interaction between flood and food supply, or drought and the stability of governments, theory becomes not a luxury of the Academy, but a necessary aid to social action.

In brief, the theory holds that in using physical resources, *people engage in behavior that combines adaptation to extreme events with both purposeful and incidental adjustment.* In doing so, they tend to move from one pattern of behavior to another across thresholds of awareness, action, and intolerance. Individual and collective choices as to what mix of adjustments is adopted reflect a kind of "bounded rationality" that is influenced by a variety of factors among which the event characteristics, human experience, resource use, and material wealth are especially important. National policies, on the other hand, tend to shift from adaptive behavior to disaster relief, to damage control, to comprehensive strategies. Exclusive dependence upon any one adjustment on the part of nation, community, or person reduces the resilience of both natural and human systems. As this is demonstrated in mounting vulnerability to catastrophe, the comprehensive strategies command greater attention from policy makers. At a critical period of transition from one behavior pattern to another, affluent countries as well as poor countries become particularly susceptible to disruption following natural extremes. With most countries in the process of transition, the environment is made more hazardous.

MODES OF COPING

Human occupance of a segment of the biosphere never starts with a clean slate: when people move into a new environment (Europeans into Australia, Africans into the Caribbean, South Asians into East Africa), they carry with them biological and cultural adaptations developed in their previous environment. Depending on the character of those adaptations, a people may be able to cope better or worse with the environment into which they move. Whatever their initial success in coping with the new circumstances, that success can be improved upon through a social learning process such as choosing land use and cropping patterns to protect the individual and the society against the prevailing adverse elements while gaining net income from the natural assets.

Slow to be aroused, human societies do, nevertheless, respond with vigor to the more extreme events. Disaster situations often call forth strong action for dealing with the emergency itself, and then preventing it from happening again—even to a radical shift either in location or in the pattern of livelihood and resource use practiced in such societies. Ways of coping are described as adaptation (biological and cultural) and adjustment (incidental and purposeful). Taken together, the many and varying coping actions can be grouped into four modes, separated by three recognizable threshold levels—*awareness, action, intolerance.* These are suggested diagrammatically in Figure 8.1, which follows the framework appearing in Chapter 2. Initial occupance (migration from elsewhere) brings with it a set of cultural and biological adaptations, which, together with incidental adjustment, serve to absorb the hazard potential of extreme events.

The first mode of coping with hazard is *loss absorption.* A society absorbs the impact of environmental extremes and remains largely unaware that it is doing so. The character of loss-absorption capacity can only be described in terms of biological and cultural adaptation and incidental adjustment.

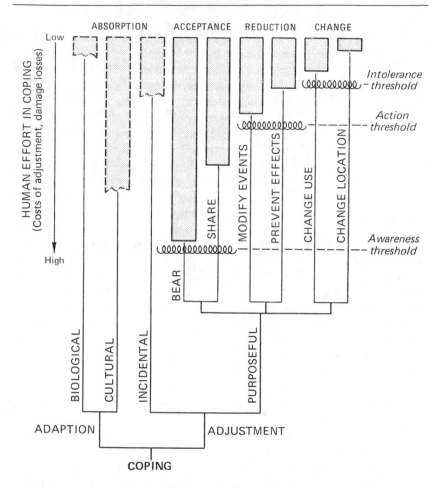

FIGURE 8.1. Modes of coping with natural hazard. Four major modes of coping with natural hazards involve loss absorption, acceptance, reduction, and change of use and livelihood. Moving from one mode to another involves crossing a significant threshold of awareness, action, or intolerance. In a rough, impressionistic global average, the adjustments are scaled in the amount of human effort committed to each.

The mode of loss absorption is separated from that of *loss acceptance* by the threshold of *awareness*. In this mode a society arranges to bear the loss, often by sharing it with a wider group than those directly affected.

Loss acceptance yields to *loss reduction* when a second threshold is crossed—that resulting from a determination to take more positive *action* to reduce loss. Yet a third threshold—*intolerance*—divides loss reduction from the mode of *radical change* in either location or resource use. This occurs when the loss is seen as no longer tolerable.

Although the four modes of coping appear universal, a steady progression through the modes is not invariably found. A society may remain in one mode with respect to floods, while crossing one or more thresholds with respect to drought. The thresholds themselves are not fixed levels, and movement in relation to them may be occasioned by a shift in any one of several variables. Changes in human response result from the cumulative individual, collective, and national choices previously described.

Absorption

In each piece of the mosaic of human use of the earth, people function within the characteristic norms of their society, exhibiting distinctive social preferences, uses of available technology, and uses of natural resources. In each situation society and environment interact to form a unique pattern of hazard exposure and of resilience to extreme events. Moreover, two societies occupying the same environment may, by virtue of their different adaptations (biological and cultural) and their incidental adjustments, be differently affected. To illustrate: small fluctuations in moisture availability—events scarcely observable by the traditional agricultural system—cause loss to commercial plantations. The extent to which a society remains unaffected by such fluctuations in nature, fluctuations that for others would be regarded as hazardous events, is its *absorptive capacity*.

Such capacity reflects the sum of customary practices and devices (such as land use) embedded in everyday livelihood that enable human societies to absorb extreme events without being made aware of significant harm. The origin and intent of those practices and devices may be lost from conscious memory. Other such practices may involve more self-conscious purpose, either because of their recent invention or obvious functionality. Whatever their origin, these become part of usual (as opposed to exceptional) human behavior and do not depend upon awareness of hazard for their continued operation.

Incidental adjustments (such as certain choices of building design or materials) also affect absorptive capacity. Unlike the practices of cultural adaptation, those of incidental adjustment are usually undertaken consciously and may be short-lived. Practices adopted for other ends may be quickly expanded or abandoned, thus strongly affecting absorptive capacity.

While absorptive capacity is a useful concept, it is not easy to define operationally or to compare analytically between societies and hazards (see Chapter 2). Thus, for example, absorptive capacity may be smaller for a modern industrial state than for a traditional society because the technology used is more subject to disruption. Conversely, the wealth of an industrialized society may serve to increase its absorptive capacity.

When extreme events in nature exceed the absorptive capacity of a system, and losses inflicted are for the first time recognized as significant, the society or some of its members cross the threshold of awareness. Movement across the awareness threshold might be occasioned also by changes in level of wealth, in technology, in social organization; by changes in resource use; or by innovations in environmental monitoring, social data collection, or accounting procedures.

The built-in absorptive capacity of a society may itself involve costs. A large number of hazard-related actions unconsciously followed are not achieved and maintained without allocation of resources. The social price of a high level of absorptive capacity may be the payment of a continual environmental "rent" not consciously recognized as such. Thus, the intercropping practiced in tropical agriculture can provide protection against drought or small moisture deficits without the society's consciously recognizing that it performs this function. Changing the cropping pattern by replacing it with monoculture in the interest of more efficient farming may also change the absorptive capacity of the system. Its increase or decrease will depend on the net balance among all factors that influence absorptive capacity.

Acceptance

The response of the victims to an extreme event that just exceeds the awareness threshold is commonly rather passive: the loss is recognized and tolerated. People prefer to bear known ills rather than take action the outcome of which may be equivocal or uncertain. Such reluctance to take further action seems to be a universal characteristic. The chronic effects of air pollution or the mild tremor of a barely perceptible earthquake are passed off as inconsequential; damage resulting from a tidal surge along the shore or the periodic flooding of a farmer's fields may be accepted with a shrug. People have a capacity for learning to live with hazard events, and, provided that the impact is not great, they may prefer to live with it than do something about it.

The pattern of *loss acceptance* does not, of course, rule out various actions to provide for the more vulnerable members of a society — the poor, the weak, the old. In the form of compensation payments, gifts, fiestas, transfers of wealth and resources, insurance, tax concessions, and the like, societies find many ways to soften the impact of loss. The more evenly the losses are spread, and the smaller the loss that each household or group has to carry, the more tolerable losses are made without further relief measures. Yet, while the acceptance of loss varies in degree, it is limited. In all societies a point arrives at which people reach the limits of their loss acceptance and cross the threshold of positive action.

Reduction

With the crossing of this action threshold, individuals or societies are galvanized into a search for more effective adjustments. In most hazardous areas the initial response is made in relation to a specific hazard event — usually a disastrous or catastrophic one. Swift emergency action follows, to be succeeded by a resolve to learn from the experience and if possible to prevent a recurrence.

This mode of response is characterized by efforts to control the hazard event itself or to reduce the vulnerability of individuals and groups. In either case the goal is *loss reduction*. The kinds of adjustment practiced have been described in Chapter 2. There is often a preference for adjustments that seek to control natural events over those requiring behavioral, social, or institutional change. The preference for "learning to live with it," characteristic of the acceptance mode, now translates into "learning to prevent it." In both cases change in the human side of the man–environment equation is avoided or put off as long as possible. The initial concentration on the modification of hazard events is commonly followed by a more mixed strategy in which ways of preventing effects are developed and adopted. For most societies dealing with most hazard situations, in most places, the actions adopted are effective in that they provide for survival and for maintenance of life to an extent beyond that possible without them.

Use/Location Change

Before the *intolerance threshold* is reached, both societies and individuals explore fully the methods of loss reduction. On rare occasions, usually only in very extreme conditions, the pressure of a hazard event forces a wholesale reappraisal of the situation. When the event does occur, the threshold of loss intolerance may be crossed, and the people involved become open to a consideration of choices that would otherwise be thought impossible. When individuals or groups exhaust their capacity for action or find the choices open to them no longer effective for a given use or location, three directions of radical change are possible: substantial changes in resource use, change in location, or a combination of these.

When migrations occur they arouse considerable interest, as in the relocation of the small population of Tristan da Cunha to Britain. On a much larger scale are the migrations in the early 1970s of starving people from the semi-arid grasslands of the Sudano–Sahelian zone to towns and cities farther south. Although in some cases such migrations will be temporary — many residents of Tristan da Cunha eventually went back — many of those migrants from the drought-stricken southern fringes of the Sahara will never return. Once relocated, they will have to begin to learn new skills in a new environment.

Usually use/location changes are associated also with institutional and social changes of other sorts. Thus, resource use may become more extensive if new lands are occupied at a lower population density than previously, or with lower capital investment per unit area. Changes from more to less intensive use and the reverse may be undertaken in response to hazard.

An urban floodplain may be preserved in an open, natural state, serve as a public park, parking lot, a residential or commercial district, or be occupied by a highly intensive damage-prone electronics plant or library. Along this continuum, shifts in use take place in the direction of either more intensive or less intensive use, and for many reasons, among which the impact of natural hazard is only one. Nevertheless, such hazard-related changes are initiated in order to reduce damage potential, or to increase the capacity to modify the hazard, or to avoid the hazard altogether.

A change in hazard exposure often accompanies a change in location. For most of the world's societies, attachment to place, especially native place, is extremely strong. While in some sense a change in resource use is interchangeable with relocation, there seems nevertheless a desire to "stay put" that exceeds the necessity to "do something different." Even in relatively mobile societies, housing, infrastructure, and industry remain stable; and for many cultures, place and use are so intertwined that a drastic move may require a change of livelihood that is difficult to contemplate even under the press of the most extreme events. Many of those swept by the cyclonic storm from Char Jabbar and similar islands in the

Ganges delta returned to the same resource use in the same location, though knowing that a recurrence of the disaster was inevitable. The inertia characterizing the use of the Wilkes-Barre floodplain is no less than that characterizing the coastal delta farmers of Bangladesh.

Although the specific reasons may differ fundamentally, the behavior seems remarkably constant from place to place. Once human settlement has been achieved and a commitment to a location has been made in terms of investment of capital and sense of affinity or identity with that place, complete abandonment rarely occurs. Organized attempts to remove people and property from floodplains usually flounder, and efforts to relocate the center of a city into a less earthquake-prone site encounter enormous resistance from the townsfolk, even though the new site may only be a few hundred meters distant. It is usually more feasible to reduce hazard loss potential by changing use than by changing location.

The modes of coping described in this theoretical analysis allow no avenue of final escape. Since there are no other modes, some loss to those occupying hazard areas is inevitable: it cannot be avoided. The loss can, however, be greater or smaller in relation to the benefits of resource use, depending upon the choice made or the mixture of adjustments adopted.

THE MIX OF ADJUSTMENTS

The mix of adjustments described in Chapter 2 is an accumulation of individual and collective effort and varies widely, as being affected by key characteristics of both the environment and the society. Each mix represents a distinctive combination of effort.

The absorption, acceptance, reduction, or evasion of hazard loss is part of the total effort required for resource use and extraction. The price of living in a risky place is twofold: the cost of adaptation and of adjustment, and the loss of life and of wealth suffered. The compensations for the risk lie in the livelihood and the welfare of the people. In a very crude way such effort can be summed up and the sums can be compared.

In global perspective, the proportion of effort expended on the different adjustments (estimated in common value terms) appears as a rough but regular (decreasing) scale. More losses are suffered by individual people than shared with the community; more effort is shared than is expended on preventing effects; more effort is directed toward changing use than toward changing location (see Figure 8.1). This global perspective breaks down at small scales, and the mix of purposeful adjustments varies widely, as has been shown among individuals (p. 52), at selected sites (p. 122), and among nations (p. 176).

Factors Affecting the Mix

On the global scale, four factors are widely found to affect the mix of human response to hazard: (1) a few characteristics of extreme events; (2) the localized experience with hazard and success of adjustment; (3) the intensity of resource use; and (4) the level of material wealth attained. These represent a convenient condensation of many variables, each factor serving as a surrogate for related or included characteristics of nature and society. Some others are excluded. Nevertheless, they account for much of the variance commonly observed in studies of human response.

Extreme Events

The individual characteristics of natural disasters affect the levels of the three thresholds described above. Such levels may be expected to be higher or lower according to the frequency, duration, areal extent, speed of onset, spatial dispersion, and temporal spacing of extreme events.

In the light of the basic distinction between intensive and pervasive natural events described in Chapter 2, the impact of a pervasive event makes an environment unacceptable and encourages evasive changes, whereas that of an intensive event can soon be forgotten, and the long interval between such events can discourage any permanent shift. That is to say, migration out of a drought-prone area such as northeast Brazil in any dry year or the Great Plains dust bowl of the 1930s is not uncommon. Migration out of earthquake-prone lands rarely occurs, even in the immediate aftermath of a disaster.

Experience

When the experience of an extreme event has been recent, the society is more likely in a state of awareness of both the threat of the event and of ways to deal with it, for such disasters help to push societies over the awareness threshold. As memory recedes into the past, however, the awareness may diminish and the experience be forgotten. Movement across any of the three thresholds into a new mode rarely takes place except in the aftermath of a damaging event. This is not to suggest that academic or theoretical learning is of no use, but only that it can be used most effectively after a disaster makes a society receptive to a new mode of response.

Resource Use

The intensive/extensive dimension may be used as a partial surrogate for the many and varied features of resource use. More intensive uses, with higher capital or labor investment per unit area, are readily alerted to the presence of a hazard, so that for them the threshold of awareness is lower. To illustrate, urban floodplain property has little absorptive capacity, and as such invites disaster, whereas a winter flood over a field during an unproductive season may not constitute a hazard at all. Small losses on an urban floodplain give rise to a loss-reduction response, whereas similar losses in agricultural areas are often tolerated more passively.

In spite of inspiring a more ready awareness of hazard and a stronger propensity to take action than does extensive use of land, intensive uses are slower to lead to use/location changes. Radical change of use, or migration elsewhere, is more likely to correspond to lower intensity use than it is to higher intensity.

Material Wealth

In the circumstance of great material wealth the awareness threshold is reached later than under conditions of poverty or little material wealth. People are less likely to suffer from small losses; the possibility of incidental adjustment is greater. The early awareness associated with little material wealth is not followed by a low threshold of action. Thus, people with little material wealth may reach the awareness threshold more quickly, but they are slow to take reductive action. Loss tolerance is lower and lasts for a shorter length of time among those of greater material wealth. Further, once the loss-reduction behavior is begun, it tends to persist longer in conditions of greater material wealth. Societies with little wealth tend to reach the threshold of intolerance sooner and have a greater propensity to change their resource use or to migrate. They have, after all, less to lose.

These four factors affecting the mix of adjustments arise from natural events, livelihood systems, and societal conditions. And although they encourage or constrain certain human responses, there is still ample choice. Since ultimately it is understanding of causes that will generate the ability to choose wisely, it might be helpful to review some prevailing conceptions about the fundamental causes of natural disaster.

THE CAUSES OF NATURAL DISASTER

An overwhelming majority of the people asked about hazard and disaster in their own localities view the occurrence as either unaccountable or as an act of nature or of God (or gods) or some other supernatural force. Rarely is it viewed as an act of people. (The distinction made in this book between extreme events in nature and human intervention in such events is not obvious to many who deal with disasters.) Three perspectives on the cause of disaster may be discerned, all differing in the emphasis that they place upon the causal factors nature, technology, and society.

Natural Science

The peasant farmer cultivating land in a flood-, cyclone-, or drought-threatened place shares with a scientific position the idea of environmental hazard as an external nature-generated phenomenon. Such hazards are, in this scientific view, essentially natural events, and as such they fall mainly within the province of natural science. As a 1972 United Nations conference report begins:

> In recent years the United Nations and its specialized agencies have noted with growing concern the problems of natural disasters which for centuries have caused great damage to human life and property, particularly in the developing countries. It is believed that not only the causes of such disasters fall within the province of science and technology, but also, in some cases their prevention, as well as the organizational arrangements made for forecasting them and reducing their impact when they occur. (United Nations Department of Economics and Social Affairs, 1972, p. 1)

This natural-science view of disaster not only puts the emphasis in scientific research on natural phenomena qua "cause" of disaster, but also introduces a bias into the choice of mitigating measures and of ways to reduce impact. Thus, among the recommendations for coping offered by the U.N. Advisory Committee on the Application of Science and Technology to Development are:

. . . [that] more effort be applied to scientific research aimed at better understanding of the phenomena and a sounder interpretation of their causes, leading to an improvement of the scientific basis of the forecasting;
. . . [that] technological research be undertaken or fostered to ensure a better protection against the effects of natural disaster; [and]
. . . [that] more action was needed to improve existing warning systems. (United Nations Department of Economics and Social Affairs, 1972, pp. 21–24)

In view of this orientation, it is not surprising that of the thirty concrete recommendations of the advisory committee almost all focus on new scientific investigations of natural disaster. Only two are directed toward human behavior.

Technology

In this orientation, attention is centered on the supposed paradox "presented in man's apparently growing susceptibility to injury from natural hazards during a period of enlarged capacity to manipulate nature" (Burton, Kates, and White, 1968). Geographical research into natural hazards began with attempts to explain the rising level of flood losses in the United States in the face of unprecedented efforts and expenditures to control them (White et al., 1958). Out of such research grew the "technological" explanation:

The prevailing public approach has been to offer immediate relief and then to turn to the technological approach. Dams follow floods, irrigation projects follow droughts.
We find that when carried out in isolation without adequate reference to social considerations, the technological adjustments may lead to an aggravation of the problem rather than an amelioration, as when upstream reservoir construction encourages increased invasion of a Tennessee Valley flood plain of Chattanooga. Commonly, the benefits received are short-run, and involve the elimination of numerous "small" losses at the cost of greater long-term losses often of a catastrophic nature. (Burton et al., 1968)

In the ensuing years since 1958, this view of a heavy-handed, technology-exacerbating hazard has been elaborated. The rise of environmental consciousness has led to critical (sometimes overcritical) assays of the link among technology, environmental deterioration, and human well-being (Farvar and Milton, 1972), and culminated in the U.N. Conference on the Human Environment. During the same period the collaborative research of the Commission on Man and the Environment of the International Geographical Union provided corroboration for the problems associated with technology in a cross-cultural context (White, 1974, pp. 11–13).

Reliance on a technological approach to hazard control has undoubtedly increased loss in many instances. It has been accompanied by a parallel orientation in research. As a review of the U.S. research effort, estimated to cost the taxpayer some $40 million per year, concluded,

> Research today concentrates largely on technologically oriented solutions to problems of natural hazards, instead of focusing equally on the social, economic, and political factors which lead to non-adoption of technological findings, or which indicate that proposed steps would not work or would only tend to perpetuate and increase the problem. In short, the all-important social, economic, and political "people" factors involved in hazards reduction have been largely ignored. They need to be examined in harmony with physical and technical factors. (White and Haas, 1975, p. 1)

Society

Social variables have been neglected both as being crucial to hazard reduction and as constituting causes of disaster loss. Both neoclassical and welfare economists have found the pattern of investment in water resource development (especially in flood control, irrigation, navigation, and power generation) to be partially responsible for disaster (Renshaw, 1957; McKean, 1958). They traced the preference for large-scale technological "solutions" to failure of the market process and to the subsequent bias, in allocation of public funds, toward the construction of large public works (Eckstein, 1958; Krutilia and Eckstein, 1958). Artificially depressed interest rates, it was argued, lead to distortion of the investment pattern, providing higher degrees of flood control or drought control (in the form of irrigation) than was justified by a rational appraisal of economic benefits and costs.

Neoclassical economists, allowing only minimal government intervention, wanted the market processes to engender more sensitive assessments of the risk run by individuals and firms. Radical critics, for their part, attribute to these self-same free-enterprise processes the exacerbation of hazard as a result of the "marginalization" of the Third World countries.

> The result of such a process is that the underdeveloped population is isolated from the traditional indigenous resource base. . . . As the

underdeveloped population attempts to discover alternative strategies of production on the edges of the imposed system that has controlled the traditional indigenous resource base, it is forced to accept strategies that contain fewer insurance or adaptive mechanisms for survival. The new strategies leave the underdeveloped population more vulnerable, more disaster prone to the vagaries of the environment; they have been forced into marginal economies of the underdeveloped countries' resource base. Marginalisation is the process which leaves the underdeveloped population more vulnerable to the vagaries of the environment. As a process, marginalization allows an explanation of the trend observed earlier that there is an increase of natural disasters in underdeveloped countries despite the fact that the probability of the natural hazards has not increased. (Baird et al., 1975)

An Interactive Explanation

Each of these perspectives on nature, technology, and society is of value; none by itself is adequate. A more nearly satisfactory explanation for the increasing disaster vulnerability of both developed and developing nations is yielded by the view that nature is neutral, and that the environmental event becomes hazardous only when it intersects with man. The event leads to disaster when (1) it is extreme in magnitude, (2) the population is very great, or (3) the human-use system is particularly vulnerable. In many cases the productive areas to inhabit are prone to landslide, flood, cyclone, earthquake, or drought. Human beings remain and prosper in such places. They might prosper even more than they do and, through a judicious selection of adjustments to hazard, they might suffer less; but some incidence of disaster is the inevitable concomitant of human use of the earth. If, however, disaster is on the increase, as the data of Chapter 1 suggest, then such increase must originate in environmental change, technological change, or social change.

ENVIRONMENTAL CHANGE

The global environment is known to be subject to drastic and dramatic change. What is still in doubt is the scale of such change: it is especially difficult to estimate change in the frequency and the magnitude of these events. For residents of a coastal town, the occasional cyclone or the rare tsunami may bring profound environmental change, and it hardly helps to be reassured by scientists that hurricanes and tsunamis constitute part of their "normal" environment. On the standard time scale of basic climatic and geological change, for the span of human time, such an environment may appear essentially changeless. The average 14 years that an Iowa farmer works a specific farm site, or even the 200 years of oral history of a clan of African farmers, appear as seconds on the climatological time scale and much less on the geological one.

Within human time scales important secular trends do develop in climate. Whether such deviations constitute climatic or geological change in the lexicon of geophysical science, they are profoundly significant to human society. Their accurate recognition is still a matter of earnest scientific controversy, even for such parameters as mean temperature. For extreme events randomly distributed in time and space, detection of trend or cycle is a matter of great uncertainty and debate. As noted in Chapter 2, the situation is compounded by the ill-defined effects of human modification of the environment. The duration and persistence of such effects cannot yet be forecast, nor is there adequate knowledge of the forces producing them. For the first time, man himself is recognized as a potential agent of change on a global scale. It had long been known that man can accelerate soil erosion, even bring on flood and drought by localized agricultural practices and water regulation. Now the possibility emerges that human action (still small compared with the energy of natural forces) might have a trigger effect: a small-scale change induced by man might cause a critical shift in the balance of atmospheric and geologic forces and bring about widespread calamity.

Although worldwide climatic and oceanographic changes figure prominently in the concerns of scientific groups at this time, other widespread perturbations—such as disruption of grassland ecosystem or urban soil–water relations, which may not show themselves for some years—are possible and deserve consideration. The chemical, biological, and radiological contamination of the biological food chains and biogeochemical cycles that lead inexorably back to man are known to be vulnerable to the scale of industrial and technological development that now exists, and will become more so as the economy continues to grow and expand. Profound disturbances in the cycling of the basic supplies of carbon, nitrogen, phosphorus, and sulfur are underway.

Individuals, groups, and societies seeking to adjust to natural hazard exposure face a situation in which the assumed "state of nature" may change in ways and at times that defy confident prediction. Man is engaged in activity that may be viewed as a game with capricious nature. Yet the effects of environmental change, while significant for the process of adjustment to natural hazard, are perhaps small compared to the rapidity of social change in many parts of the world, and hence often underestimated. In learning to cope with possibly massive man-made alterations in the state of nature, people may benefit from observing how the stress of societal change affects their response to natural extremes.

SOCIETAL CHANGE

As nature constrains the modes of coping, so do the wealth, technological capacity, and organization of society. These modes vary by societal type, as shown in Figure 8.2. Thus, in a traditional or folk society, coping is characterized by a high absorptive capacity—a large number of adjustments widely shared among individuals and communities. These adjustments often involve patterns of behavior or of agricultural practices more cooperative with nature than controlling it. The adjustments typically are low in per capita cost and may often be added to in small increments, although the aggregate expenditure by any one household may represent a substantial share of its capital investment. To this extent, adjustments are flexible—easily increased or reduced in scale; they are also closely related to social customs and supported by behavior norms and community sanctions. Total technological and capital requirements are commonly low.

Governments in pre-industrial societies generally do not find it possible to protect their citizens against the vagaries of nature: the web of individual and group adaptations does so in part, and political agencies respond in rudimentary ways. In tribal societies, government beyond the local level is usually poorly developed; national action serves chiefly to offer a little relief when conditions are acute. The more highly organized societies of China or the Islamic world, and central or national governments, have from time to time played a significant role in the construction, maintenance, and operation of irrigation and flood-control systems—notably in certain river-based civilizations. Yet, whether a strong central authority, with its overtones of despotism (Wittfogel, 1957), has been necessary to provide the social organization and discipline for the maintenance of complex systems of environmental management remains in doubt. Some irrigation programs have prospered in a rather casual way, and extensive operation of large-scale works in Egypt and China continued even when the governments were weak.

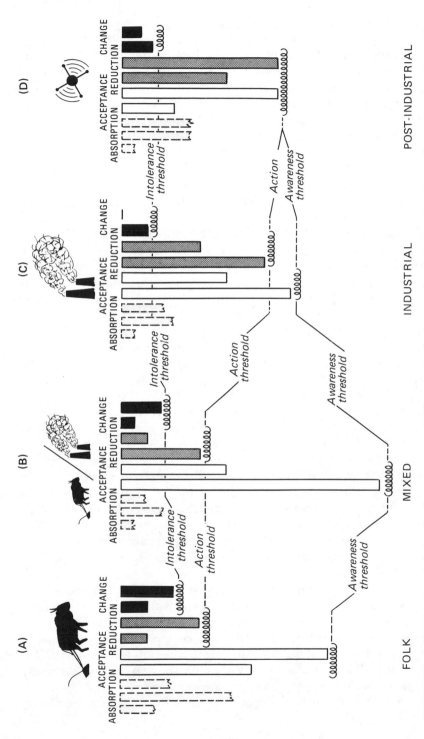

FIGURE 8.2. Modes of coping, by development stage. Differences in the four modes of coping, the mix of adjustments, and threshold levels are shown proportionately to the amount of effort (adjustment cost and damage loss) in four archetypal societies.

235

Except for intervention in the great irrigation systems, national governments in these societies have played a largely passive role.

The pattern of folk response is a large number of adjustments and a high rate of adoptions among individuals and communities. They often involve modifications of behavior or of agricultural practices more cooperative with nature than controlling it. Typically the adjustments are low in per capita cost and may often be added to in small increments, although the aggregate expenditure by any one peasant may represent a substantial share of his capital investment. To this extent adjustments are flexible—easily increased or reduced in scale. They are closely related to social customs and supported by behavior norms and community sanctions. Technological and capital requirements are commonly low.

While this coping pattern requires cooperative action by community or local groups, it does not depend on outside assistance in the form of technical knowledge, financial assistance, or governmental approval. The pattern may vary drastically over short distances according to local differences in hazard severity or cultural practice. Though effective in preventing loss of property and loss of life from low-magnitude hazard events, the folk pattern is ineffective in the prevention of catastrophic events. When such occur, massive government or social intervention may be needed if greater disaster or wholesale migration is to be averted.

In the modern industrial state a different pattern emerges. Acceptance shifts from bearing loss personally, or sharing with kin, to sharing with the wider society by means of relief or insurance. The growing technological capacity to manage or manipulate the environment encourages the reduction of hazard by emphasizing policies for the control of nature. These favored adjustments require interlocking and interdependent social organization, and they tend to be uniform in application, inflexible, and difficult to change. The construction of dams, irrigation systems, or seawalls, and the design of monitoring, forecasting, and warning systems with complex equipment would be clearly beyond the scope of individual action.

All developing countries are composed of both folk societies and others that are tied into the sophisticated world economy. In such mixed populaces the more traditional adjustments—individual and community—are displaced, and national governments attempt to take up the slack, despite limited resources of funds and skilled manpower. The more effectively they fill the gap, the more rapidly do the traditional coping modes atrophy.

In the postindustrial society, the modes of coping are slowly evolving like the peoples themselves. There is a marked shift to a broader coping pattern, a turning away from reliance on the control or modification of nature toward a comprehensive policy for lowering of the damage potential and a concern with the effects of disaster policies on the well-being of the people involved.

Common to almost all models of societal change is a focus on one-direction development, with an explicit or implicit assumption of growing complexity. An economy of broadening scale and deeper specialization replaces simple structure, and there is an increasingly complex integration of function at different levels. Growth in economic activity is a concomitant of political stability (or attempts to impose such stability) but seems to involve a decline in the integrative social bonds of family and community life. Specialization may be seen as leading to either increased skills or pauperization on the part of the labor force; hierarchical integration can foster either organization or imperialism or supranational cooperation; new social relations can find expression either in greater personal freedom or in alienation.

As societies enlarge their scale of action, differentiate their functions, and integrate, their adjustment to hazard changes. As folk societies (Figure 8.2A) mix with industrial societies (Figure 8.2B), the number and importance of their cultural adaptations and incidental adjustments diminishes. The absorptive capacity of a folk society to prevent loss lessens when new livelihood systems are introduced. As these systems come into being, the appropriate new skills of hazard coping are slowly learned: how to cope with an extreme event in a new setting or with a new source of livelihood is knowledge that is acquired late (Figure 8.2C).

If the resulting loss to a society exceeds its tolerance for average or for catastrophic disruption, that society will turn to radical changes in use or location. Coastal hurricane zones, for example, will be evacuated permanently; citrus growers will convert to a less risky crop. As the disruptive impact of the extreme event becomes more acute, a postindustrial society will shift toward a broader mode of coping (Figure 8.2D), or make a basic change in use and/or location. In so doing, the society seems to rediscover the advantages of diversity and flexibility inherent in the folk society.

These differences in modes of coping lead to different social costs for each of the societal types. These are shown in Figure 8.3 and effectively divide the world into the high damage losses and costs of industrialized societies and the proportionally high damage and loss of life of developing folk and mixed societies. All these costs may be enlarged by the rapidity of social change.

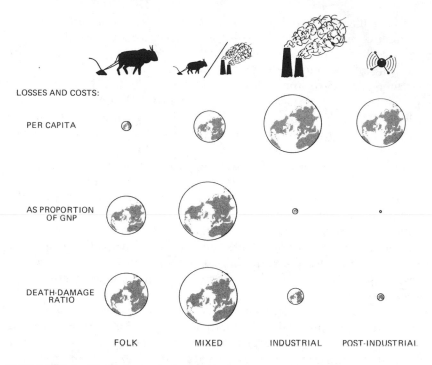

FIGURE 8.3. Hazard costs, by societal type. Absolute losses from natural hazards are much greater in industrialized than in developing societies, but proportional losses and deaths are extremely high in the mixed and folk societies.

IS A LESS HAZARDOUS ENVIRONMENT ATTAINABLE?

In the light of the foregoing analysis, it will be appreciated that a reduction of disaster potential cannot be achieved easily and that further increase in the incidence of disaster is the probable "wave" of the next decade. The U.N. Disaster Relief Office has sought to identify the more disaster-prone developing nations in order to give them priority ranking for technical assistance in early warning and disaster prevention. According to the theory advanced in our book, neither the poorest nations nor the slowest-developing ones are considered disaster-prone. The high-hazard nations would be those favored with natural resources that are undergoing change in use.

A comparison of the death rate from disaster (excluding drought) with GNP per capita (Table 8.1) suggests that the "least developed" nations are less disaster-prone than those with somewhat higher per capita income. These poorest among the developing

TABLE 8.1. Comparative Death Rates from Natural Disasters, by National Income

Annual death rate (per million pop.)	Per capita national income			
	Least developed ≤ $200	Less developed $200–$1,000	Industrial ≥ $1,000	Total
High > 200	2	14	1	17
Moderate and low ≤ 200	10	24	19	53
Total	12	38	20	70

Source: Based on data from Dworkin (1974).

countries are isolated, land-locked, and often have low resource potential and low population density. If they suffer less the intensive hazard associated with coastal or floodplain location, they are more vulnerable to drought and soil erosion; in any case, total losses are small in absolute terms, although proportionately high.

In Europe a greater absorptive capacity for most hazards would be expected by reason of the greater age of its settlement and longer experience of its people. Such capacity would come from adjustments intended to modify the extreme events and prevent their effects. In contrast, Australia, Canada, and the United States make use of their extensive territory and social mobility to change location and land use more readily. On balance, lower per capita losses would be expected for population at risk in Europe than in the larger North American and Oceanic nations.

On a global scale, an increase in material wealth and a more equitable distribution of that wealth would lead toward reliance on modification of events and prevention of effects. Although the absorptive capacity of the global system might be high, the acceptance of loss, once recognized, would probably diminish with the growth of material wealth. Intensifying the hierarchical organization and specialization would reinforce the trend toward seeking technological solutions for hazard problems. There may have to be repeated on a world scale the same sort of slow learning process as that set in motion by human encounter with hazard on a local scale.

As integration of the global system proceeds and as pervasive global hazards emerge, the thrust toward loss acceptance and then loss-reduction measures will be strong. While the U.N. Disaster Relief Office will help spread hazard losses more widely, it will not, of course, produce anything approaching equity; it can only buy time or delay the need for further action by means of increasing loss acceptance. An even stronger thrust is likely toward loss-reduction measures. These may include — and clearly have done so in certain instances — ambitious plans for technological solution, ranging from traditional irrigation schemes to watering the Sahara with desalinated water from industrial plants powered by nuclear generating stations. The settlement of human populations on floating, manmade islands, or in submerged cities on the seabed, is among the science-fiction solutions advocated in all seriousness by well-intentioned if over-imaginative scientists and engineers. From the perspective of hazard theory, such fantasies are not likely to help much in the short run. If seriously entertained, in the long run they may well exacerbate a situation of growing severity.

BY WAY OF SUMMARY . . .

Nature, technology, and society interact to generate vulnerability and resilience vis-à-vis disaster. In the short run the global toll in damage will continue to rise, while loss of life will be reduced substantially.

The long-term thrust of development in nations is toward reducing the social cost of hazard to society, but in periods of rapid transition societies become peculiarly vulnerable to hazard. A central task for international cooperation should be to ease these transitions.

Hazard vulnerability varies among nations, emphasizing loss of life in the developing, and catastrophic damage in the highly industrialized.

Loss of property is most likely to grow in rapidly developing societies, particularly those with coastal locations and those exhibiting large inequalities of wealth. Loss is more likely to diminish in industrial nations encouraging a wide range of adjustments, and in developing nations with widespread sharing and self-reliance. With large numbers of nations in transition, the environment for the short term will continue to become more hazardous. The forces propelling the world toward more and greater disasters will continue to outweigh by a wide margin the forces promoting a wise choice of adjustments to hazard. There is hope for a safer environment, but it cannot be achieved easily or soon.

CHAPTER NINE

Emerging Synthesis

By 1992, four important new developments in natural hazard research and applications had emerged: the scientific research itself had vastly expanded in reach and analysis; an ambitious international program of research and application to reduce disasters was launched; the exceptional challenge of global environmental change became the newest focus for research and policy; and sustainable development became a new criterion for adjustment to hazard. Taken together, these promise to lay the basis for a new synthesis that should be possible by the end of the century.

THE REACH OF RESEARCH

The environment as hazard has always been broader than the compass of this volume. At the time of publication of the first edition of this book, hazard research and its influence on policy was already moving beyond the set of natural physical hazards, the focus on empirical experience of extreme events, and the adjustment responses of individuals and national governments. Current research includes biological and technological hazards as well as physical hazards, seeks to integrate the extreme with everyday experience, asks basic theoretical and generic questions, and uses social as well as individual categories of analysis. It continues to be problem-oriented, theoretically eclectic, socially active, aversive to professional conflict, and thoroughly interdisciplinary. Mitchell (1989), O'Riordan (1986), Whyte (1986), Drabek (1986), Palm (1990), and Alexander (1991) have extensively reviewed these developments. Hazards continue to be a good lens through which to advance the understanding of the nexus of nature, society and technology (Kates, 1987).

The post–World War II development of hazard research was stimulated by the publication in 1945 of White's *Human Adjustment to*

Floods. Between that date and the publication of *The Environment as Hazard*, hazard research focused on a set of geophysical natural hazards characterized by extreme events, their perception, and the adjustment responses of both individuals and national governments, initially in North America but eventually in other industrialized and developing countries. The basic theoretical model that underlined this work viewed the impacts of natural hazards as derived from the interaction between nature and society and strongly affected by feedback in the form of human adjustment. Considerable attention also was given to the process of choice in adjustment, using as a basic theoretical model Simon's (1956) view of bounded rationality.

In recent years several important commentaries on this work have been published (Waddell, 1977; Torry, 1979; Walker, 1979; Hewitt, 1983; Watts, 1983a; Emel and Peet, 1989; Palm, 1990). They project two broad sets of criticisms. One set is critical of the use of extreme geophysical events as a starting point for analysis, noting the value of alternative units of analysis such as places, livelihood systems, social groups, or societies. This focus on extreme events, it is asserted, detracts from comprehending the human ecology of everyday existence, perhaps implying a simplistic determinism. In real life, the everyday and the extraordinary are intermixed. Collectivities of hazards — natural, social, and technological — threaten human existence, the salience of any one threat changing over time. Moreover, the vulnerabilities of people are rooted in the precariousness of everyday existence as well as in the rare and extreme event. Adjustments that are viewed as intentional responses to previously experienced extreme events do not adequately portray the subtlety and complexity of interacting social, livelihood, and natural systems.

A second critique of early hazard research finds it insensitive to the social and economic constraints that serve to limit the choice of both people and governments. The widespread use of choice and decision models and the focus on such psychological constructs as individual perception implied that humans are masters of their fate to a much greater extent than is valid. This emphasis on choice of adjustment seemed to ignore the reality of the constraints and structure that are part of social existence. It focused on the isolated decision, often ignoring the nested structures of society and their dynamics that bind and constrain the decision process. Such an emphasis appeared to be rooted in an ideology of individualism that ignored the reality of what was, at best, a constraint, and at worst, the oppression that characterized everyday social experience, especially in the developing world.

There is merit in these major critiques and we see them as part of the continuing scientific debate (Burton, Kates, and White, 1981; Torry, 1986). Indeed, they demonstrate how hazard research is an open-ended scientific inquiry into one aspect of the human condition and its potential betterment. Its critics are as much a part of the evolving research process as its early innovators. Within such a view, the hazard "paradigm" is not static and needs no defense against a Kuhnian scientific revolution but should undergo revision through substantial scientific research. This has happened in the 15 years since the first edition was written, and, as is so often the case in productive science, the published critique has lagged behind the actual research.

Hazard research has given stronger emphasis to some previously underdeveloped ideas: encompassing slow and cumulative events, exploring biological and technological dimensions, emphasizing complexity, social causation, and extracting theories of the middle range. At the same time, it has become more eclectic and interdisciplinary, and thus somewhat less coherent.

From Sudden Events to the Slow and Cumulative

If, as a matter of convenience, the hazard domain is divided into three sectors — natural, technological, and social (mindful that pure forms of each are fiction) — then the literature through 1978 is dominated by a very small sliver of that domain, namely, geophysical hazards. Even within that portion of the hazard domain, slow cumulative hazards such as soil erosion or climate change were poorly studied, partly due to the bias toward more dramatic and sudden extreme events. This changed significantly with the preparation by geographers of basic scientific papers for the U.N. Conference on Desertification. The paper on population, society, and desertification (Kates, Johnson, and Johnson-Haring, 1977) extended the hazard paradigm to study the slow, cumulative spread and intensification of the desert-like conditions directly affecting some 50 million people around the world. The consequences of coastal erosion were likewise the subject of systematic review (National Research Council, 1990).

Working from a neo-Marxist perspective, Blaikie (1985) examined the hazards of soil erosion in the developing world, and together with Brookfield (Blaikie and Brookfield, 1987), put together a series of studies on the social causes of land degradation. In the Amazon, Hecht, Anderson, and May (1988) examined the causes and consequences of deforestation. Similar studies emerged from within development anthropology and cultural ecology (Little and Horowitz, 1987). These pioneering studies support the widely held view (Leonard, 1989; Durning, 1989) that impoverished people and degraded environments are strongly related. A recent review of some 30 case studies of the links among hunger, poverty, and environment found extensive support for a spiral of household impoverishment and environmental degradation (Kates and Haarmann, 1992).

As the slow but steady losses of soil or vegetation were seen as prominent hazards of the land (McKell, 1989), even slower, more cumulative shifts in the atmosphere became major foci of hazards research. Since 1978, a strong consensus developed that widespread climate change forced by the accumulation of greenhouse gasses in the atmosphere is underway, yet its magnitude, rates of change, and regional effects remain highly uncertain (Houghton, Jenkins, and Ephramus, 1990). Hazard research provided a ready approach for studying potential impacts of climate change and societal adjustment to it. A new field of climate impact assessment (Chen, Boulding, and Schneider, 1983; Kates, 1985a; Kates, Ausubel, and Berberian, 1985) developed, and exemplary studies of agriculture (Warrick, 1980; Liverman, Terjung, Hayes, and Mearns, 1986; Warrick and Gifford, 1986; Parry, Carter, and Konjin, 1988; Parry, 1990; Riebsame, Changnon, and Karl, 1991), and the regional economies of the central United States (Rosenberg and Crosson, 1991), Mexico (Liverman, 1990, 1992), and the United Kingdom (U.K. Climate Change Impacts Review Group, 1991) were produced. Within the traditional set of natural geophysical hazards, the pervasive, slow, and cumulative hazards came to be the latest major research frontier and were studied increasingly.

From the Geophysical to the Biological

Neglected within the traditional set of hazard studies were the biological natural hazards, such as infectious diseases or pests. This was true despite the considerable evidence of the importance of biological hazards to human existence and security. Beginning in 1982, an International Geographical Union Working Group on Environmental Perception initiated a collaboration that included natural hazard geographers, cultural ecologists, and biologists interested in integrated pest management. Three international meetings, jointly sponsored by the UNESCO Man and the Biosphere Program, the International Organization for Biological Control, and the IGU Working Group on the Perception of Pests and Pesticides, placed major emphasis on developing countries. Studies focussed on general fact finding through the construction of national profiles of pesticide use, success stories of integrated pest management, studies of perception and adjustment to pests by smallholder farmers, observations of pesticide use, and the recon-struction of pesticide flows between industrialized and developing countries, including their return link in the form of residues on imported products (Tait, 1981; Tait and Napompeth, 1987). Gold-man's (1991) study of the pests of three major crops in Kenya and farmers' perception and response to the hazard exemplifies this broadening of perspective.

The most pervasive of all hazards is disease, but infectious disease has been little studied within the hazard framework. A major exception was the work on water-related diseases in studies of domestic water use (White, Bradley, and White, 1972). This began to change with the reoccurrence of new forms of infectious diseases (AIDS, other sexually transmitted diseases, Legionnaires' Disease, and toxic shock syndrome) that serve as sources of surprise (Kates, 1985) and fear (Kirby, 1990) in society. In the context of under-standing hunger, the interactions between infectious disease (mea-sles, malaria, diarrhea) and malnutrition, particularly in children, came under increasing scrutiny (Wisner, 1989; Millman and Kates, 1990).

In a close parallel to the hazard framework, a broader ecological view of disease has been developed following the pioneering work of Selye (1956, 1974). Health is seen not simply as the absence of disease, but as a state of physical and mental well-being that is subject to the total environment and all sources of stress (Howe 1980; Burton, 1990; Foster, 1992). In this per-spective, hazards are addressed in the total context of experience in specific locations. This has been applied especially in the rapidly growing urban environments of developing countries (Burton and White, 1983).

From the Natural to the Technological

Although pests and infectious organisms are major biological threats, pesticides constitute a technological hazard, and the new diseases are often transmitted by technology. The broadest outreach of hazard research has been to technological hazard, the beginnings of which were explored in the context of the air pollution studies in Mexico and the United Kingdom (Chapter 3) and were being more actively pursued as the first edition of this book was being published (Kates, 1977a, 1977b, 1978; Whyte and Burton, 1980). Hazard research, then, merges with the field of risk assessment both through conscious examination of the similarities and differences (White, 1988a; Kasperson and Pijawka, 1985; Pijawka, Cuthbertson, and Olson, 1988) and by the research undertaken (Burton, Whyte, and Victor, 1983).

The natural hazard paradigm inspired a set of related studies of technological hazards: a broad survey of the range of hazards and responses to them; an estimate of costs and losses, surveys of scientific and lay perceptions of technological hazards, development of an expanded model of hazard causality, and a series of detailed case studies and policy analyses (Burton, Fowle, and McCullough, 1982; Kates, Hohenemser, and Kasperson, 1985). The work was interdisciplinary (Starr, 1972; Lowrence, 1976; Lawless, 1977) and benefitted from new skills and insights as well as the substantial experience with natural hazards. It was comparative from the beginning, contrasting hazards and responses to them and seeking measures of the magnitude of their consequences.

A comparative study of 93 hazards yields a basic taxonomy that identifies five classes of extreme hazards: intentional biocides (e.g., chain saws, antibiotics, and vaccines), persistent pathogens (e.g., uranium mining and rubber manufacture), rare catastrophes (e.g., natural gas explosions and commercial aviation crashes), common killers (e.g., auto crashes and coal mining) and diffuse global threats (e.g., CO_2 release and supersonic aircraft). A few hazards evidence multiple extremes (e.g., nuclear war, recombinant DNA, and pesticides) (Hohenemser, Kates, and Slovic, 1983). The scientific indicators used to create this classification were also used to examine laypersons' perception (Slovic, Fischoff, and Lichtenstein, 1985). Laypeople in many ways order hazards similarly to the scientific judgement, but they betray strong biases with respect to a few hazards and tend to compress the scale of differences, overvaluing small hazards and underestimating large hazards as judged by the scientific literature.

Estimates of costs and losses from technological hazards were made (using data from the late 1970s) and are contrasted with data on natural hazards including biological as well as geophysical. These are shown in Table 9.1.

The causal model of technological hazard expands on the classical hazard distinction between events and their consequences, being more precise about causation and questioning the uses and employment of technology as well as its potential for hazard. The

TABLE 9.1. Comparative Hazard Sources in the United States and Developing Countries

| | Principal causal agent[a] | | | |
| | Natural[b] | | Technological[c] | |
	Social cost[d] (% of GNP)	Mortality (% of total)	Social cost[d] (% of GNP)	Mortality (% of total)
United States	2-4	3-5	5-15	15-25
Developing countries	15-40[e]	10-25	n.a.[f]	n.a.[f]

[a]Nature and technology are both implicated in most hazards. The division that is made here is made by principal causal agent which, particularly for natural hazards, can usually be identified unambiguously.
[b]Consists of geophysical events (floods, drought, tropical cyclones, earthquakes, and soil erosion); organisms that attack crops, forests, livestock; and bacteria and viruses which infect humans. In the United States the social cost of each of these sources is roughly equal.
[c]Based on a broad definition of technological causation, as discussed in source.
[d]Social costs include property damage, losses of productivity from illness and death, and the costs of control adjustments for preventing damage, mitigating consequences, or sharing losses.
[e]Excludes estimates of productivity loss by illness, disablement, or death.
[f]Not available, as no systematic study of technological hazards in developing countries is known to us, but we expect them to approach or exceed U.S. levels in heavily urbanized areas.
Source: Harris, Hohenemser, and Kates (1985).

model serves to identify the theoretical range of adjustment and recognizes the interventions that can prevent or mitigate harm (Hohenemser, Kasperson, and Kates, 1985). Case studies explored in detail the processes of hazard causation and management (Kasperson, Kates, and Hohenemser, 1985). A major undertaking evaluated equity considerations between places and generations (Kasperson, Derr, and Kates, 1983; Kasperson, 1983) and social groups such as workers and the general public (Derr, Goble, Kasperson, and Kates, 1985).

Natural hazard research melded with the field of risk assessment (Kates, 1978; Whyte and Burton, 1980; Kasperson and Kates, 1983; Pushchak and Burton, 1983; Covello and Mumpower, 1985; Mitchell, 1989; Smith, 1992). Yet this melding has spurred less reciprocity than might have been expected by researchers and managers concerned with reducing the threat of technological hazards (White, 1988a; Kirby, 1990). Indeed, the research traditions flowed like two streams in roughly parallel courses in an alluvial valley—touching here or there, joining each other during high water, but for the most part separate, with few direct connections. At a conceptual level, natural hazard research and risk assessment share a similar structure, but in the overwhelming proportion of the literature this is obscured by the diverse subject matter and policy issues. They draw from different scientific disciplines, but whereas geophysicists study earthquakes and biologists and chemists study toxic substances, the study of how those phenomena are perceived and how society may choose to evaluate them are common themes. Observations from natural hazards research of importance to risk assessors include the following:

- Almost all hazards have both natural and technological components, with the mix differing from one place to another.
- An analysis of risk needs to take account of how it is perceived by the people directly affected, individuals and organizations involved in responding to risk, as well as the perceptions of scientific and technical analysts.
- Unless a risk analysis includes the social structure within which individual decisions are made, it may fail to clarify either the process or the consequences of those decisions.
- Risk communication differs by the nature of the message, the channels used to communicate it, and the varied circumstances of age, gender, income, education, and experience in which people subject to risk find themselves.
- The use of benefit–cost analysis to appraise the efficacy of proposed methods of handling risk has severe limitations and may be misleading rather than helpful in providing tools for decision.
- The separation of analysis of risk from appraisal of risk management may be administratively convenient or seemingly more objective, but such neatness may lead to unrealistic findings as to action options and social response.
- There is great need for postaudits of risk assessment and management that might be helpful in further research and mitigation.

From Extreme Events to Complex Conditions

At the heart of comparative studies of technological hazards is a common definition equivalent to that of the extreme event in natural hazards: the release of energy or materials that threaten humans or what they value. Studying the temporal and spatial patterns of release of energy and materials suggests a continuum between the rare, sudden, massive release of energy and materials, as in a nuclear power plant accident, and the continuing chronic release of energy and materials, as in urban air pollution. And many hazardous phenomena seem to combine elements of both acute and chronic events and of natural and technological events, as when an inversion traps air pollutants over a city and prevents their dispersal.

Everyday life and language harbor even larger, more complex hazardous conditions such as hunger or poverty, disease or ill health (Douglas and Wildavsky, 1982). In life we live them and as language we use them both as threatening events or conditions and as consequences of other, often multiple-causative events. For one of these complex conditions, *hunger*, faculty at Brown University created a conceptual model that describes the causes, consequences, and interrelationship among three conditions of hunger: individual food deprivation, household food poverty, and regional food shortage (Millman and Kates, 1990).

As Figure 9.1 shows, individual *food deprivation* is what doctors and nutritionists define as hunger: the failure in a person to take in, absorb, or retain sufficient nutrients to maintain health, provide growth, and undertake activity. Individual deprivation at the level of the household becomes *food poverty* in which a household is unable to provide sufficient food to meet the nutritional needs of its members either by its own production, by exchange for its labor or products, or by other entitlements (Sen, 1982). Similarly at the level

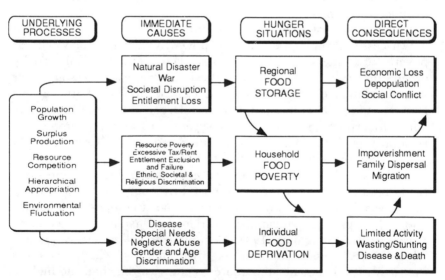

FIGURE 9.1. Hunger: A cascade of hazards.

of a region, hunger becomes *food shortage,* the absence of food within a region sufficient to meet the needs of its population, such shortages in their most extreme form being what are commonly called famines (Currey and Hugo, 1984).

Thus, hunger spans conditions that range from famine to the deprivation of a needed micronutrient — vitamin A — in small children. Its immediate causes range from natural hazards such as drought or flood to gender discrimination in households. And its consequences include death and disease in people and impoverishment in households and regions. From the macroregion to the microperson, a cascade of troubles flows downward in time of food shortages and upward in the aggregation of consequences.

From the Individual to the Social Group

Throughout this volume, differential hazard vulnerability is discussed mainly in broad and aggregate terms between particular countries, between developing and industrialized societies, and between societies in transition. But much of the analysis of loss and adjustment focuses on the conditions of individuals and their households. As this volume was first being published, studies were under way to examine more closely the differential vulnerability to hazard within countries, between communities, and within communities — vulnerability conditioned by social group, economic class, or livelihood system.

In the first of such empirical studies, Wisner (1977) showed how access to drought adjustments in Kenya varied by the wealth or poverty of farmers and their access to land. This was followed by other studies that examined how the lack of access to resources or wealth diminished the ability of households to cope with various hazards and how vulnerability was enlarged for some groups by being incorporated into a world system of capitalist production relationships (O'Keefe, Wisner, and Baird, 1977; Garcia, 1981; Watts, 1983b; Hewitt, 1983; Scott, 1984).

The belief that hazard vulnerability was intimately and positively related to poverty was supported by analyses that suggested that it was the poorest countries that were most vulnerable to hazard and had the greatest per capita hazard losses (Susman, O'Keefe, and Wisner, 1983; Hagman, 1984; Wijkman and Timberlake, 1986). This differed somewhat from our own analysis, which suggested that it was poor but not the poorest countries that had the highest losses. Poor countries were especially vulnerable to hazards because of the development taking place — development that, for the most part, had not yet occurred in the poorest countries. In part, these divergences in interpretation arose because of somewhat different time series that included or excluded drought, which is difficult to measure and to report over time (Kates, 1980). Continuing drought in Africa falls heavily on the poorest, least-developed countries.

In contrast to the earlier focus in natural hazards on individual choice, studies of technological hazards began with social groups. For example, a major study of risk perception, although using individual survey forms, compared League of Women Voters members, a businessmen's club, students, and risk experts (Slovic, Fischhoff, and Lichtenstein, 1985). Another major set of studies in the technological domain focussed on equity considerations, the fairness of hazard exposure and consequences by place, by generation, and by social group. Case studies of radioactive waste (Kasperson, 1983), temporary workers in the nuclear power industry (Melville, 1981), lead in the atmosphere and workplace (Hattis, Goble, and Ashford, 1982), and field application of parathion (Johnson, 1982) highlighted this work along with a penetrating look at the ethical standards available for the resolution of equity issues posed by technological hazard (Derr,Goble, Kasperson, and Kates, 1981). More recently, hidden hazards have been identified — hazards hidden because of the position, ideology, or values of the social groups affected (Kasperson and Kasperson, 1991).

With the strong neo-Marxist emphasis on the primacy of economic relationships, there was unfortunately less study of the differential vulnerability to hazard related to other powerful social dividers: age, gender, social status, and ethnicity. Yet the studies on Bangladesh reported in this volume demonstrate that women, children, and the poor were the prime victims of tropical cyclone. Too few studies have examined these differential causes of vulnerability and how they affect mitigation, emergency, and reconstruction activities (Rivers, 1982; Bolin and Bolton, 1986).

In general, studies of differential vulnerability to natural hazards have been strong on societal critique and weak on practicable prescription. However, in a recent study, Downing (1991) uses differential group vulnerability to suggest ways to improve the famine early warning and response system in African countries. To sharpen the warnings and to identify groups requiring special attention he combines available indicators of wealth, livelihood source, and the special needs for food of women and children.

In addressing technological hazards, research on group, place, and generational equity readily lends itself to such policy prescriptions as law and regulation designed to share the risks of waste facilities (Kasperson, 1983) or to diminish the inequity that allows risks in the workplace hundreds of times higher than in the general environment (Derr et al., 1981).

From the Empirical to the Theoretical

Hazard research had its origins in the descriptive empiricism of geography, the philosophical pragmatism that characterized an earlier American social science. There has, however, been a growing consideration of theory that has taken three general forms: the transfer of existing grand theory, selective eclecticism, and distinctive hazard theories of the middle range (Kates, 1988).

An example of the transfer of grand theories is found in the debate over mortality in the Sahelian drought. In this case, theories of modernization contrasted with theories of underdevelopment. They projected diametrically opposite estimates of loss of life during the droughts of 1968–1973 and 1910–1915 (Caldwell, 1977; Berry, Campbell, and Emker, 1977; Kates, 1981). In contrast, selective eclectics draw from many sources, and this can be exemplified by efforts such as the recent work of Palm (1990), who draws from theories of political economy, culture, social organization, and decision making to create an integrative framework for study and exploration, and of Alexander (1991) to provide a framework for research and teaching. Similarly, the work on technological hazards sought a generic synthesis between biological and physical law, risk perception, and the social management of hazard (Kates, Hohenemser, and Kasperson, 1985). Equity analysis tried to integrate empirical research and a searching application of fundamental canons of justice (Derr et al., 1981).

Our preference has been to propose and explore theories of the middle range that purport to explain important aspects of hazard-related behavior observed empirically or address distinctive qualities of the intersection of nature, society, and technology. Several of the key ideas described in this volume seem to have stood the short test of time:

- Hazards and resources are uniquely related; people encounter hazard in the search for the useful.
- Nature, technology, and society interact to generate vulnerability and resilience to hazard. Thus, there are no *uniquely* natural, social, or technological hazards, nor can hazard consequences be meaningfully examined separate from

human response. As the hazards of our times seem to become more numerous, more extensive in space and time, and more complex in cause and consequence, so do these interactions become more apparent.

- The long-term thrust of development in nations is towards reducing the social cost of hazard to society — but in periods of rapid transition, societies become peculiarly vulnerable to hazard.

Subsequent to their development in this volume, these key ideas have been borne out by much of the work on technological hazards as well as tragically validated by disaster in Seveso, Bhopal, Chernobyl, and Mexico City. Studies of hazards over time have suggested additional theoretical ideas related to lessening, catastrophe, and amplification.

The "lessening" hypothesis (Bowden, Kates, Kay, Riebsame, Warrick, Johnson, Gould, and Wiener, 1981) states that continuing societies facing recurrent hazards adjust to them more successfully, so that similar levels of extreme events elicit a lessened toll of losses, damages, and costs. But by actions taken to lessen hazard impacts, societies may make themselves catastrophically vulnerable to extreme events that exceed the expected.

Thus, a related hypothesis known as the "catastrophic" hypothesis was developed. The most common example of such behavior was derived from the early observations of the catastrophic potential losses from levees on floodplains. Levees successfully prevented small and moderate floods but trapped the waters of great floods that overtopped the levees and thus made such floods more damaging. Historical studies of the recurrent drought on the Great Plains (Warrick, 1980; Warrick and Bowden, 1981) and in the Sahel (Kates, 1981) have been used to explore these two hypotheses. The analysis was extended in time with a computer simulation of the Tigris–Euphrates flood plain over the last 6,000 years (Johnson and Gould, 1984) and was further extended in studies of three other long-term reconstructions in the Nile Valley, the Central Mayan Lowlands, and the Basin of Mexico (Whitmore, Turner, Johnson, Kates, and Gottschang, 1990).

Lessening can be readily seen in Figures 1.3 and 1.4, where despite the rising number of large natural disasters, deaths have diminished drastically. But there is little support for a catastrophic hypothesis that would predict fewer but larger disasters. These figures use deaths as a metric, and different conclusions emerge from the more limited studies of property losses. Not only have such property losses increased dramatically (Figure 1.7), but in the face of lessening, the long-term costs of adjustment surely increase as well.

Finally, the newness of technological hazards and the observed limitation in their perception and comprehension by both scientists and laypersons (Slovic et al., 1985) has recently spawned both theoretical and empirical work on the social amplification of risk (Kasperson et al., 1988). This conceptual view of risk focuses not on personal experience, but on risk depicted by communication sources, including scientists, news media, activist social organizations, informal networks of friends and neighbors, and public agencies. These information "stations" can intensify or weaken signals of risk and filter or reinforce certain attributes. Such amplified portrayals of risk events then ripple through society, affecting other parties, distant locations, or even future generations (Kasperson and Kasperson, 1991).

THE INTERNATIONAL DECADE FOR NATURAL DISASTER REDUCTION

During the 1980s support grew for the establishment under the United Nations of an International Decade for Natural Disaster Reduction. The concern stemmed from increasingly detailed journalistic coverage of the great disasters of the time and from rising confidence on the part of natural scientists and engineers that their research contributions could reverse the rising flow of human losses from natural events (National Research Council, 1987).

Without comprehensive appraisal of prior attempts to cope with losses from extreme events in nature or of the lessons to be learned from the conduct of international decade programs, a new Decade was launched. Its aims were stated by the United Nations General Assembly:

> The objective of this Decade is to reduce through concerted international actions, especially in developing countries, loss of life, property damage and social and economic disruption caused by natural disasters . . . and . . . its goals are: (a) To improve the capacity of each country to mitigate the effects of natural disasters expeditiously and effectively, paying special attention to assisting developing countries in the establishment, when needed, of early warning systems; (b) To devise appropriate guidelines and strategies for applying existing knowledge, taking into account the cultural and economic diversity among nations; (c) To foster scientific and engineering endeavors

aimed at closing critical gaps in knowledge in order to reduce loss of life and property; (d) To disseminate existing and new information related to measures for the assessment, prediction, prevention and mitigation of natural disasters; (e) To develop measures for the assessment, prediction, prevention and mitigation of natural disasters through programs of technical assistance and technology transfer, demonstration projects, and education and training, tailored to specific hazards and locations, and to evaluate the effectiveness of those programs. (UN, General Assembly Resolution 42/169, 1987)

The disasters envisioned were primarily those caused by short-duration events. The methods to be followed in achieving those goals were described only in general terms. The manner of incorporating national government efforts into the global venture was not specified. Thus, in its early stages the Decade was largely a time of exploring the feasibility and possible effectiveness of options for action. These options seemed heavily weighted on natural science research and engineering technology (Mitchell, 1988; Bunin, 1989; Housner, 1989; Dynes, 1990). The difficulties encountered during those opening years revealed in striking fashion the theoretical and practical obstacles encountered in many areas where government policy turns to action designed to reduce vulnerability to extreme events in nature. As described by a committee of the U.S. National Research Council (1987), the Decade effort in the U.S. was intended to

provide knowledge and mitigation practices that could cut impacts of natural hazards at least 50 percent by the year 2000. Achieving this national goal requires a major program of research, technological development, project applications, and public information activities: a nationwide assessment of natural hazards and their risks; collection, analysis, and dissemination of information on hazards; an assessment of current knowledge and practices and identification of gaps in knowledge; a research program to fill those gaps; effective educational programs; and cooperative research activities in and among all relevant disciplines and professions. (National Research Council, 1987, p. 4)

This remains a commendable goal, but there is no scientific analysis to support the expectation of a 50 percent reduction in impacts by the year 2000. Improved flood and hurricane forecasting will not necessarily reduce damages drastically, even though it usually curbs loss of life. Better designs for earthquake-resistant buildings will not change the vulnerability to shock of old buildings requiring demolition or retrofitting. Thus, a country might well regard a decade in which average losses did not rise, let alone diminish, as a success.

The experience with earlier UN development decades and with the International Decade for Drinking Water and Sanitation had made it abundantly clear that unless scientific research and technological development are strongly linked with studies and action focused on human behavior and social organization, the benefits to people might be greatly or wholly lacking. The most careful system for supplying water to a village may have little impact on health unless accompanied by change in household use of water and disposal of waste (White, Bradley, and White, 1972; Burton, 1977). Similarly, forecasts and attendant warnings are only as effective as the response to them. Improved building designs are only as beneficial as their application. Maps of areas vulnerable to landslides are only as helpful as the social will to curb building in critical zones. We stress these realities because they illustrate a few of the basic lessons to be derived from natural hazards research, and they show that basing public reliance on disaster reduction solely through technical measures may be counterproductive. Failure to face up to hard facts about disaster mitigation may encourage public bodies to think they have taken beneficial action when they have ignored essential measures on the social side.

Whether or not the Decade will prove on balance to achieve its aims remains to be shown as it unfolds and as governments choose their measures to advance it. And, just as with numerous national programs, the data on losses and costs from which the relative success of the effort might be gauged are so far lacking.

THE CHALLENGE OF GLOBAL ENVIRONMENTAL CHANGE

The newest challenge for hazard research is global human-induced environmental change (Malone, 1986; Burton and Timmerman, 1989; Kates et al., 1990; Jacobson and Price, 1991; Stern, Young, and Druckman, 1992). The magnitude of such change has been documented recently through an international collaborative effort (Turner et al., 1990). A sampling of their findings shows that since the dawn of agriculture 10,000 years ago, an area the size of the continental United States has been deforested by human effort (Kates, Clark, and Turner, 1990). Today, half of the ecosystems of the ice-free lands of the earth have been modified, managed, or utilized by people. The flows of materials and energy that are removed from their natural settings or synthesized now rival the flows of such materials within nature itself. Water in an amount greater than the contents of Lake Huron is withdrawn each year for human use and is contaminated on a massive scale. All told, the Earth Transformed Project was able to reconstruct human-induced change in 13 worldwide measures of chemical flow, land cover, and biotic diversity. If we consider the total amount of human-induced change represented by these 13 measures over the last 10,000 years, then most of it has been extraordinarily recent, with 7 of the measures having doubled after 1950.

The rapidity of these changes lends credence to current fears for the earth's fate. These fears are enhanced by projections of changes yet to come, for it is over the next 60–80 years that the international forecasts and projections foresee a doubling of the earth's population. To meet their needs, some project a fourfold increase in agriculture, a sixfold rise in energy use, and an eightfold increase in the value of the global economy (Anderberg, 1989). The environmental changes caused by such enormous intensification of production and consumption could well be too much for maintenance of existing quality of human health and well-being and for the life-support systems of the earth.

The threats to the earth are many and varied. Adopting criteria that consider which are most likely to occur, to cause the most harm, or to affect the most people, three groups of large-scale and long-term threats emerge. First are the *global atmospheric* concerns of nuclear winter, stratospheric ozone depletion, and climatic warming from greenhouse gases. Second are massive *assaults on the biota,* specifically, deforestation in the tropical and mountain lands, desertification in the drylands, soil degradation in humid or irrigated lands, and species extinction, particularly in the tropics. Last are the large-scale introductions of *pollutants:* acid rain in the atmosphere, heavy metals accumulating in soils, and contaminants in surface and ground water. We choose the most perplexing of these, the case of global warming, to explore this challenge in detail.

Warming and Its Consequences

Global warming had been raised periodically as a matter of concern since Arrhenius in 1890, but major scientific interest only developed in the wake of reports on the rapid increase in carbon dioxide in the atmosphere and likelihood of subsequent global warming (Bolin, Döös, Warrick, and Jäger, 1986; National Research Council, 1987). This research eventually led to the organization of a major intergovernmental and international scientific review by government agencies in the Intergovernmental Panel on Climate Change (IPCC).

One report of the IPCC (Houghton, Jenkins, and Ephraums, 1990) confirmed that in the view of a prevailing body of scientific opinion among atmospheric scientists, the global mean surface air temperature increased by $0.3°$ to $0.6°$ Celsius over the preceding 100 years. They projected a further increase of $2°$ to $5°$ Celsius some time in the next century if present trends of increase in the human-induced "greenhouse gasses" (carbon dioxide [CO_2], methane [CH_4], chlorofluorocarbons [CFCs], and nitrous oxide [N_2O]) reached a level of twice that of pre-industrial CO_2.

Despite the confidence expressed in the IPCC reports about the validity of the theory of global warming, it is acknowledged that considerable uncertainty remains concerning the rate of warming, the actual amounts, and its distribution around the world (Wigley and Raper, 1992). On the scientific side, for example, the role of the oceans as a carbon sink is poorly understood. As concentrations of CO_2 increase, the absorption rate in the oceans may accelerate. Similarly, the role of cloud cover is an insufficiently known phenomenon. If cloud cover increases with warmer surface temperatures, this could increase the reflectance (albedo) of the atmosphere and so moderate greenhouse warming but, by reducing nighttime cooling, could reinforce it positively. There are other possible positive feedback effects, such as the thawing out of large areas of permafrost in the tundra and the subsequent release of huge quantities of methane, a powerful greenhouse gas.

Major uncertainty also concerns the likely regional effects of global warming. Projections of warming are based on several quasi-independent general circulation models of the atmosphere, which support the general conclusions but vary in regional detail. Temperature and precipitation increases are likely to be much higher in some regions than others, but there is insufficient detail within models or agreement between models to predict the quantities. The general expectation is that temperature increases will probably be greater toward the poles and significantly less toward the tropical and equatorial regions. Temperatures may increase to a greater extent in continental interiors than in coastal zones. There is much less agreement on the patterns of precipitation. Most models predict sea-level change under the doubled CO_2 scenario, threatening to some degree low islands and delta regions of coastal countries (Warrick and Farmer, 1990; Warrick, Barrow, and Wigley, 1991).

A Complex Hazard

Within the context of natural hazard research, global warming is a complex threat that overlaps the traditional concepts of hazards and their sources, combining the natural and the technological, the geophysical and the biological, and the everyday and the extraordinary. Humankind has necessarily adapted to climate change as a natural phenomenon over the millennia (Ausubel, 1991). But prospective global warming superimposes an anthropogenic change on top of a naturally changing phenomenon. Indeed, the melding of the two processes is such that it is currently impracticable to separate them. The distinction between geophysical hazards and biological hazards, and the early focus on the former to the neglect of the latter, may become even more evident as the nature of the threat of global warming becomes apparent. Depending upon the pace of climate change, global warming can be expected to stimulate significant biological and ecological changes.

Within the range of geophysical hazards, natural hazard research was concerned to a large extent with extreme events and more recently with the slow, pervasive, and cumulative hazards. In the case of climate change we have the emergence of a complex hazard phenomenon that is slow and cumulative but may also include a greater frequency of extreme weather events. Climate change is projected by the general circulation models in terms of long-term incremental changes in global and regional means. The actual manner of change as it is likely to be experienced will be in the distribution of weather events, especially extreme weather events. Heat waves and droughts might be more common with rising temperatures, floods and snowstorms with increased precipitation.

Adjusting to Global Warming

Despite these substantial areas of uncertainty, the threat was perceived as significantly great that an International Framework Convention on Climate Change (United Nations General Assembly, 1992) was signed by 152 nations in Rio de Janeiro in 1992. It will be followed by more specific protocols at a later date to reduce specific greenhouse gas emissions and to adapt to global warming. The framework convention was modeled in part upon the Vienna Convention on substances that deplete the ozone layer. That convention laid down a broad framework under which it was possible to negotiate the Montreal Protocol providing for specific scheduled reduction and eventual elimination of CFC production.

However, it will not be easy to be optimistic about the prospects for more precise protocols. The difficulties stem from the nature of the hazard itself; the complexities of the required adjustment process, and the supposed economic costs of adjustment compared with the benefits to be gained in terms of damages prevented (Nordhaus, 1991). What has been learned from hazards research that can now be brought to bear on the problems of responding to global warming?

An important contribution of natural hazard research has been its clarification of the concept of adjustment, and in particular the distinction between those adjustments that involve attempts to control or modify the extreme events and those intended to reduce individual and societal vulnerability to them. This distinction has emerged as poles in the research and policy debate over responding to the threat of global warming (Committee on Science, Engineering, and Public Policy 1991). One pole favors *limitation* (also *mitigation* or *prevention*), where limitation refers to slowing or reducing the rate of emissions of radiatively active greenhouse gases. The other pole speaks of *adaptation* to the new climate conditions, where adaptation refers to actions that may be taken to reduce the harmful impacts or take advantage of the beneficial opportunities of climate change.

Frequently posed as either/or options, natural hazard research asserts that a socially optimal response is likely to involve a mixture of both types of response, and in any event, some efforts will be undertaken to do both. As with structural adjustments to floods, limiting emissions of carbon and the other greenhouse gases may prove very costly beyond an initial set of low-cost opportunities. On the other hand, adaptation also implies serious social costs, albeit less easily observed and measured. Moreover, the ability to limit or to adapt is not distributed equitably among or within countries. Poor countries, places, and groups have few options open to them compared to wealthy countries or classes.

The time and space scales of adjustment to global warming are unprecedented both in the slow, cumulative, temporal expression of the hazard (albeit punctuated by sudden and extreme events), by the global but disparate regional spatial expression, the long lead times available and required to change human livelihood systems, and the collectivity of will required to manage a truly global resource (Clark, 1985, 1989; Burton and Cohen, 1992). Creating such common purpose is difficult because global warming currently involves an unequal distribution of costs and benefits. This distribution will change with time as some of the larger developing countries increase their own contribution to greenhouse gas emissions. There will be winners and losers. Countries that are prepared to make sacrifices in the global interest are likely to be critical of those who do little or nothing.

To encourage international response and to achieve distributional justice, a new range of adjustments enters into consideration. These include such items as technology transfer to the developing countries and other concessionary arrangements to enable developing countries to play their part in the limitation of greenhouse gas emissions without compromising their development efforts. Similar adjustments are required for adaptation to climate change. These might include, for example, technical and financial assistance to developing countries to enable them to reduce their vulnerability to the changing environment. Negotiations for a framework convention led to the creation of specific provisions of this kind. More difficult to address will be the plight of the impoverished and marginalized people *within* countries. These indeed may be the ultimate victims of global warming, lacking the buffers to adaptation to the hazards of new climatic regimes or the resources to take advantage of them.

Finally, there is the complex issue of adjusting to the adjustments themselves — that is, the adjustments to the high social costs implied by some responses to global warming. In the current climate debate, major emphasis is being placed on the need to limit the emissions of greenhouse gases. The sources of these emissions, especially carbon dioxide from the burning of fossil fuels, are so pervasive and such an integral component in the modern industrial economy that serious attempts to reduce them may have widespread repercussions. Similarly, efforts to get developing countries to limit their utilization of fossil fuels or the production of related products such as fertilizers can have profound effects. These repercussions will themselves constitute a hazard to many users, and will need to be a source of adjustment activity. Indeed, the priority question in industrialized societies may not be how societies adjust or adapt to climate change, but rather how they adjust or adapt to the new policies, prices, and technologies that will be created and implemented to deal with anticipated change. An important part of the challenge of global warming will be to put decision making with respect to the hazard in its full social context.

Given the cloud of scientific uncertainty about the future magnitude and rate of change, there is resistance in some policy quarters to action and a pronounced tendency to do little until uncertainty is reduced. This policy runs the risk of producing greater losses in the longer run because by the time the uncertainty is reduced it will be extremely difficult or too late to prevent major change, and the cost of taking the necessary steps will increase. There is the risk of doing too little too late or of doing too much too soon. Here in the debate the precautionary principle enters. Decisions as to what actions would minimize future regret are difficult, but they are implicit in any policy and cannot be separated from consideration of the implications for any policy of its likely effects upon the maintenance for future generations of the basic resources of land, water, air, and biota.

Sustainability: A New Criterion

In dealing with global warming, natural hazards research is obliged to broaden its paradigm once more to take account of the emergence of a complex global hazard, and to do so without losing the incisiveness that has brought new insight and practical value in situations where people confront the hazards of nature and poverty on a daily basis. Increasingly, it is becoming clear that in the long run measures to deal with extreme events in nature need to be considered in the framework of sustainable development. They need to be seen as one important component of social policy and program assuring that the resources of an area will be maintained undiminished for future generations while providing support for a satisfactory quality of life of present and future communities. Such a synthesis is beginning to emerge (Burton and Kates, 1986).

Since the publication of the Brundtland Report, *Our Common Future*, by the World Commission on Environment and Development (1987), the concept of sustainable development (Clark and Munn, 1986) has become a guiding theme for environmental policy at the international level and at the national level in many countries (White, 1990). Although not an operational concept that can be precisely defined and applied (Brown, Hanson, Liverman, and Merideth, 1987), it serves a useful purpose in making explicit the indissoluble link between environmental protection and equity. According to the philosophy of the Brundtland Commission, which has now been widely accepted at least in principle by many governments, the management of threats to the global environment cannot be achieved without simultaneously addressing issues of equity and development.

When we completed our comparative study of natural hazards in 1978, we proposed that all future development projects under the United Nations Development Programme be examined to assure that proper consideration was given to possible implications for hazards mitigation or response. That suggestion was parallel to the growth of interest in environmental impact statements. It did not take hold. It is as relevant in 1992 as in 1978 but in the new conditions should be reformulated, beyond the achievement of economic hazard reduction, to assurance that neither extreme events nor response to them detract from achieving sustainable development.

Public agencies may be expected to move away from focus on natural hazard reduction programs toward incorporating hazard mitigation as an integral part of programs for sustainable development. The test, for example, of a proposed investment in flood warning or seismic retrofitting of vulnerable buildings or design of coastal residential construction will not be solely in terms of how the costs relate to prospective benefits from reducing losses of property and life. A test will consider the implications for the sustained use of the area to meet quality needs. This will involve more explicit analysis of the extent to which occurrence of extreme events impairs the capacity of a resource use system to survive without degrading the environmental base, cultivates resilience, and reduces vulnerability. Such a test will also seek to eliminate the likelihood that an effort to make an area free from a hazard will encourage use that proves to be destructive in the long term.

Under this view, the elementary and all-embracing hazard is that the resource use will be nonsustainable. And the lasting adjustment and the ultimate adaptation will be to strike the balance between the needs of humankind and the variations in nature such that both people and their environment become less hazardous to each other.

National and Comparative Studies

● Flood   ○ Drought   ■ Hurricane   ❑ Air pollution
▲ Avalanche   △ Insurance   ▼ General

Location of comparative studies undertaken under the auspices of the Commission on Man and Environment of the International Geographical Union.

264

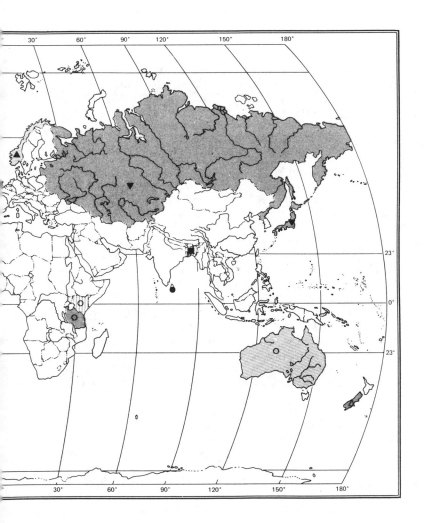

265

References

Adams, R. L. A. *Weather Information, and Outdoor Recreation Decisions: A Case Study of the New England Beach Trip.* Ph.D. dissertation, Clark University, 1971.

Ahearn, F. L., and R. E. Cohen. *Disasters and Mental Health: An Annotated Bibliography.* Rockville, MD: U.S. Center for Mental Health Studies of Emergencies, 1984.

Alexander, D. Natural Disasters: A Framework for Research and Teaching, *Disasters, 15*(3) (1991), 209–226.

Algermissen, S. T., J. Dewey, and W. A. Rinehart. *A Study of Earthquake Losses in the San Francisco Bay Area: Data and Analysis.* Report prepared for Office of Emergency Preparedness, National Oceanic and Atmospheric Administration. Washington, DC: U.S. Department of Commerce, 1972.

Allen, W. *The African Husbandman.* London: Oliver and Boyd, 1965.

Anderberg, S. A Conventional Wisdom Scenario for Global Population, Energy, and Agriculture 1975–2075. In F. L. Toth, E. Hizsnyik, and W. C. Clark, eds., *Scenarios of Socioeconomic Development for Studies of Global Environmental Change: A Critical Review.* RR-89-4 Laxenberg, Austria: International Institute for Applied Systems Analysis, 1989, pp. 201–229.

Archer, P. E. *The Urban Snow Hazard: A Case Study of the Perception of Adjustments to, and Wage Salary Losses Suffered from Snowfall in the City of Toronto During the Winter of 1967–1968.* M.A. thesis, University of Toronto, Department of Geography, 1970.

Arsdol, M. D. Van, G. Sabagh, and F. Alexander. Reality and the Perception of Environmental Hazards. *Journal of Health and Human Behavior, 5* (1964), 144–153.

Auliciems, A., and I. Burton. Trends in Smoke Concentrations Before and After Clean Air Act of 1956. *Atmospheric Environment, 7* (1973), 1063–1070.

Ausubel, J. Does Climate Still Matter? *Nature, 350* (1991), 649–652.

Badolato, E. V., J. Bleiweis, J. D. Craig, and H. W. Fleming, Jr. *Hurricane Hugo: Lessons Learned in Energy Emergency Preparedness.* Clemson, SC: Clemson University, Strom Thurmond Institute of Government and Public Affairs, 1990.

Baird, A., P. O'Keefe, K. Westgate, and B. Wisner. *Towards an Explanation and Reduction of Disaster Proneness.* Occasional Paper 11. Bradford, UK: University of Bradford Disaster Research Unit, 1975.

Baker, E. J. *Evacuation Decisionmaking and Public Response in Hurricane Hugo in South Carolina.* Quick Response Report No. 44. Boulder, CO: University of Colorado, Natural Hazards Research and Applications Information Center, 1990.

Barrows, J. Forest Fire Research for Environmental Protection. *Journal of Forestry,* 69(1) (1971), 17–20.

Bauer, B. Personal communication with authors. 1972.

Baumann, D. D., and C. Russell, eds. *Urban Snow Hazard: Economic and Social Implications.* Water Resources Center Research Report No. 37. Urbana, IL: University of Illinois, 1971.

Baumann, D. D., and J. H. Sims. The Tornado Threat: Coping Styles of the North and South. *Science, 176* (June 30, 1972), 1386–1392.

Baumann, D. D. Human Response to the Hurricane. In G. F. White, ed., *Natural Hazards: Local, National, Global.* New York: Oxford University Press, 1974, pp. 25–30.

Baumann, N., and R. Emmer. *Flood Insurance and Community Planning.* Natural Hazards Research Working Paper (NHRWP) No. 29. Boulder, CO: University of Colorado, Institute of Behavioral Science, 1976.

BCEOM (Bureau Central des Etudes pour les Equipements d'Outre-Mer). *Inventaire des Zones Inondables.* Paris: Ministère de L'Equipement et du Logement, n.d.

Becker, S. Personal communication with authors. 1974.

Berry, L., D. J. Campbell, and I. Emker. Trends in Man-Land Interaction in the West African Sahel. In David Dalby, R. S. Harrison-Church, and F. Bezzaz, eds., *Drought in Africa 2.* London: International Africa Institute, 1971, pp. 83–91.

Beyer, J. Personal communication with authors. 1969.

Blaikie, P. *The Political Economy of Soil Erosion in Developing Countries.* London: Longman, 1985.

Blaikie, P., and H. Brookfield, eds. *Land Degradation and Society.* London: Methuen, 1987.

Blair, J. P. Home to Tristan da Cunha. *National Geographic, 125* (January 1964), 60–81.

Bolin, B., B. R. Döös, R. A. Warrick, and J. Jäger, eds. *The Greenhouse Effect, Climatic Change and Ecosystems.* SCOPE Report No. 29. Chichester, England: John Wiley and Sons, 1986.

Bolin, R. and P. Bolton. *Race, Religion, and Ethinicity in Disaster Recovery.* Monograph No. 42. Boulder, CO: University of Colorado, Natural Hazards Research and Applications Informations Center, 1986.

Bowden, M. J., R. W. Kates, P. A. Kay, W. E. Riebsame, R. A. Warrick, D. L. Johnson, H. A. Gould, and D. Wiener. The Effect of Climate Fluctuations on Human Populations: Two Hypotheses. In T. M. Wigley, M. J. Ingram, and G. Farmer, eds., *Climate and History: Studies in Past Climates and Their Impact on Man.* Cambridge, England: Cambridge University Press, 1981.

Brooks, C. E. P., and J. Glasspoole. *British Floods and Droughts.* London: Ernest Benn Limited, 1928.

Brown, B. J., M. E. Hanson, D. M. Liverman, and R. W. Merideth, Jr. Global Sustainability: Toward Definition. *Environmental Management, 11*(6) (1987), 713–719.

Bryson, R. A. A Perspective on Climatic Change. *Science, 184* (1974), 753–760.

Bunin, J. Incorporating Ecological Concerns into the IDNDR. *Natural Hazards Observer, 14*(2) (1989), 4–5.

Burton, I. Types of Agricultural Occupance of Flood Plains in the United States.

Department of Geography Research Paper No. 75. Chicago: University of Chicago, 1962.

Burton, I. A Preliminary Report on Flood Damage Reduction. *Geographical Bulletin,* 7(3) (1965), 161–185.

Burton, I. Safe Water for All. *Natural Resources Forum* 1 (1977), 95–110.

Burton, I. Human Dimensions of Global Change: Toward a Research Agenda. In N. J. Rosenberg, W. E. Easterling, P. R. Crossen, and J. Darmstadter, eds., *Greenhouse Warming: Abatement and Adaptation.* Washington, DC: Resources for the Future. 1989, pp. 159–165.

Burton, I. Factors in Urban Stress. *Journal of Sociology and Social Welfare,* 17(1) (March, 1990), 79–92.

Burton, I., and A. Auliciems. Air Pollution in Toronto. In W. R. D. Sewell and I. Burton, eds., *Perceptions and Attitudes in Resources Management.* Ottawa, Ontario: Information Canada, 1972, pp. 71–80.

Burton, I. and S. J. Cohen. *Adapting to Global Warming: Regional Options.* Paper presented at the International Conference on Impacts of Climate Variations and Sustainable Development in Semi-Arid Regions. Forteleza, Brazil, February 1992.

Burton, I., C. D. Fowle, and R. S. McCullough. *Living with Risk: Environmental Risk Assessment in Canada.* Toronto, Canada: University of Toronto, Institute for Environmental Studies, 1982.

Burton, I. and K. Hewitt. *The Hazardousness of a Place: A Regional Ecology of Damaging Events.* Department of Geography Research Paper No. 6. Toronto: University of Toronto Press, 1971.

Burton, I., and R. W. Kates. The Perception of Natural Hazards in Resource Management. *Natural Resources Journal, 3* (January 1964), 412–441.

Burton, I., and R. W. Kates. The Great Climacteric, 1798–2048: The Transition to a Just and Sustainable Human Environment. In R. W. Kates and I. Burton, eds., *Geography, Resources, and Environment,* Volume 2, *Themes from the Work of Gilbert F. White.* Chicago: University of Chicago Press, 1986, pp. 339–360.

Burton, I., R. W. Kates, and G. F. White. *The Human Ecology of Extreme Geophysical Events.* Department of Geography, Natural Hazards Research Working Paper No. 1. Toronto: University of Toronto, 1968.

Burton, I., R. W. Kates, and G. F. White. *The Environment as Hazard.* New York: Oxford University Press, 1978.

Burton, I., R. W. Kates, and G. F. White. The Future of Hazard Research: A Reply to William I. Torry, *Canadian Geographer, 25*(3) (1981), 286–289.

Burton, I., and K. Moon. *The Perception of the Hazardousness of a Place: A Comparative Study of Five Natural Hazards in London, Ontario.* Paper presented at UNESCO Natural Hazards Seminar, Gödöllö, Hungary, August 1971.

Burton, I., and P. Timmerman. Human Dimensions of Global Change: A Review of Responsibilities and Opportunities. *International Social Science Journal, 41*(3) (August, 1989), 297–313.

Burton, I. and R. White. *Approaches to the Study of the Environmental Implications of Contemporary Urbanization.* MAB Technical Notes No. 14. Paris: UNESCO, 1983.

Burton, I., A. V. Whyte, and P. Victor. *The Missisaugua Evacuation: Final Report to the Solicitor-General.* Ontario, Canada: Ministry of Solicitor-General, Toronto, Ontario, 1983.

Burton, I., D. Billingsley, M. Blacksell, V. Chapman, A. V. Kirkby, L. Foster, and G. Wall. Public Response to a Successful Air Pollution Control Program. In J. Taylor, ed., *Climatic Resources.* London: David and Charles, 1974, pp. 173–191.

Byers, H. R., and R. Braham. *Methods of Estimating the Maximum Flood.* Chicago: University of Chicago, Department of Meteorology, 1959.

Caldwell, J. C. Demographic Aspects of Drought: An Examination of the African Drought of 1970–1974. In D. Dalby, R. S. H. Church, and F. Bezzaz, eds., *Drought in Africa 2.* London: International African Institute, 1977, pp. 93–100.

Carrigan, P. H., Jr., and H. G. Golden. *Optimizing Information Transfer in a Stream Gaging Network.* Water Resources Investigations Report No. 30–75. Washington, DC: U.S. Geological Survey, 1975, 30–75.

Chen, L. C., ed. *Disaster in Bangladesh: Health Crises in a Developing Nation.* New York: Oxford University Press, 1973.

Chen, R. S., E. Boulding, and S. H. Schneider, eds. *Social Science Research and Climate Change.* Dordrecht, Netherlands: D. Reidel, 1983.

Clark, W. C. Scales of Climate Impacts, *Climatic Change, 7* (1985), 5–27.

Clark, W. C. The Human Dimensions of Global Environmental Change. In Committee on Global Change, *Towards an Understanding of Global Change.* Washington: National Academy Press, 1989, pp. 134–200.

Clark, W. C., and R. T. Munn, eds. *Sustainable Development of the Biosphere.* Cambridge, England: Cambridge University Press, 1986.

Cochran, A. *Annotated Bibliography on Natural Hazards.* Natural Hazards Research Working Paper No. 22. Boulder, CO: University of Colorado, Institute of Behavioral Science, 1972.

Cochrane, H. C., J. E. Haas, M. S. Bowden, and R. W. Kates. *Social Science Perspectives on the Coming San Francisco Earthquake: Economic Impact, Prediction, and Reconstruction.* Natural Hazards Research Working Paper No. 25. Boulder, CO: University of Colorado, Institute of Behavioral Science, 1974.

Committee on Public Engineering. *Policy Perspectives on Benefit-Risk Decision Making.* Washington, DC: National Academy of Engineering, 1972.

Committee on Science, Engineering, and Public Policy. *Policy Implications of Greenhouse Warming: Mitigation, Adaptation, and the Science Base.* Washington, DC: National Academy Press, 1991.

Covello, V. T., and J. Mumpower. Risk Analysis and Risk Management, *Risk Analysis, 5*(2) (1985), 103–120.

Cuny, F. C. *Disasters and Development.* New York: Oxford University Press, 1983.

Currey, B., and G. Hugo, *Famine as a Geographical Phenomenon.* Dordrecht, Netherlands: D. Reidel, 1984.

Curry, L. The Climatic Resources of Intensive Grassland Farming: The Waikato, New Zealand. *Geographical Review, 52* (1962), 174–194.

Cyert, R., and J. March. *A Behavioral Theory of the Firm.* Englewood Cliffs, NJ: Prentice-Hall, 1963.

Derr, P., R. Gable, R. E. Kasperson, and R. W. Kates. Worker/Public Protection: The Double Standard. *Environment, 23*(7) (1981), 6–15, 31–36.

Derr, P., R. Goble., R. E. Kasperson, and R. W. Kates. Protecting Workers, Protecting Publics: The Ethics of Differential Protection. In C. Whipple and V. T. Covello, eds., *Risk Analysis in the Private Sector.* New York: Plenum Press, 1985, 257–269.

Doane, R. R. *World Balance Sheet.* New York: Harper, 1957.

Douglas, M. and A. Wildavsky. *Risk and Culture.* Berkeley: University of California Press, 1982.

Downing, T. E. *Assessing Socioeconomic Vulnerability to Famine: Frameworks, Concepts, and Applications.* Alan Shawn Feinstein World Hunger Program Research Report 91-1. Providence, RI: Brown University, 1991.

Downing, T. E., and R. W. Kates. The International Response to the Threat of Chloroflurocarbon to Atmospheric Ozone. *American Economic Review, 72*(2) (1982), 267–272.

Drabek, T. E., and J. E. Haas. *Understanding Complex Organizations.* Dubuque, Iowa: William C. Brown, 1974.

Drabek, T. K. *Human System Response to Disaster: An Inventory of Sociologic Findings.* New York: Springer Verlag, 1986.

Dupree, H., and W. Roder. Coping with Drought in a Preindustrial, Preliterate Farm Society. In G. F. White, ed., *Natural Hazards: Local, National, Global.* New York: Oxford University Press, 1974, pp. 115–119.

Durning, A. B. *Poverty and the Environment: Reversing the Downward Spiral.* Worldwatch Paper No. 92. Washington, DC: Worldwatch Institute, 1989.

Dworkin, J. *Global Trends in Natural Disasters 1947–1973.* Natural Hazards Research Working Paper No. 26. Boulder, CO: University of Colorado, Institute of Behavioral Science, 1974.

Dynes, R. R. Social Concerns and the IDNDR, *Natural Hazards Observer, 14*(3) (1990), 7.

Eckstein, O. *Water Resource Development: The Economics of Project Evaluation.* Cambridge, MA: Harvard University Press, 1958.

Emel, J., and R. Peet. Resource Management and Natural Hazards. In R. Peet and N. Thrift, eds., *New Models in Geography.* London: Unwin Hyman, 1989, pp. 49–76.

Farvar, T. M., and J. P. Milton, eds. *The Careless Technology: Ecology and International Development.* Garden City, NY: Natural History Press, 1972.

Foster, H. D. *Disaster Planning: The Preservation of Life and Property.* New York: Springer Verlag, 1980.

Foster, H. D. *Health, Disease, & the Environment.* London: Belhaven Press, 1992.

Foster, L. T. *The Development of Smokeless Zone Concept.* Paper presented to the International Geographical Union, 22nd International Geographical Congress, Commission on Man and Environment, No. 31, Calgary, Alberta, June 1972.

Frank, N. L., and S. A. Husain. The Deadliest Cyclone in History. *Bulletin American Meteorological Society, 52*(6) (1971), 438–444.

Friedman, D. G. *The Storm Surge along the Gulf and South Atlantic Coast Lines.* Hartford, CT: The Travelers Insurance Company, 1971. Unpublished.

Garcia, R. V. *Drought and Man: The 1972 Case History,* Vol. 1. *Nature Pleads Not Guilty.* Oxford, England: Pergamon Press, 1981.

Geipel, R. *Long-term Consequences of Disasters: The Reconstruction of Friuli, Italy in its International Context, 1976–1988.* New York: Springer Verlag, 1991.

Gerasimov, I. P., and T. V. Zvonkova. Natural Hazards in the Territory of the USSR: Study, Control and Warning. In G. F. White, ed., *Natural Hazards: Local, National, Global.* New York: Oxford University Press, 1974, pp. 243–254.

Glickman, T. S. *Study of Trends in Disasters, 1945–1986.* Washington, DC: Resources for the Future, 1991.

Glickman, T. S., D. Golding, and E. D. Silverman. *Acts of God and Acts of Man: Recent Trends in Natural Disasters and Major Industrial Accidents.* Washington, DC: Resources for the Future, 1992.

Golant, S., and I. Burton. *The Meaning of a Hazard. Application of the Semantic Differential.* National Hazards Working Paper No. 7. Toronto: University of Toronto, Department of Geography, 1969.

Goldman, A. Tradition and Change in Postharvest Pest Management in Kenya, *Agriculture and Human Values*, Winter–Spring (1991), 99–113.

Green, C. H., S. Tunstall, J. P. Emery, and P. I. Aggrey. *Evaluating the Non-Monetary Impacts of Flooding.* Report No. 123. Enfield, England: Middlesex Polytechnic, Flood Hazard Research Centre, 1988

Green, C. H., D. J. Parker, P. Thomson, and E. C. Penning-Rowsell. *Indirect Losses from Urban Flooding: An Analytical Framework.* Publication No. 52. Enfield, England: Middlesex Polytechnic, Flood Hazard Research Centre, 1983.

Green, C. H. and E. C. Penning-Rowsell. *Flooding and the Quantification of "Intangibles."* Publication No. 124. Enfield, England: Middlesex Polytechnic, Flood Hazard Research Centre, 1988.

Greene, M. F. Forest Fire Insurance. In W. G. Wahlenberg, ed., *A Guide to Loblolly and Slash Pine Plantation Management in Southeastern U.S.A.* Report No. 14, Macon, Ga.: Georgia Forest Research Council, 1965, pp. 281–283.

Grigg, N. S. and E. C. Vlachos, eds. *Drought Water Management.* Fort Collins, CO: Colorado State University, International School for Water Resources, 1990.

Haas, J. E., R. W. Kates, and M. J. Bowden, eds. *Reconstruction Following Disaster.* Cambridge, MA and London: M.I.T. Press, 1977.

Hagman, G. *Prevention Better than Cure. Report on Human and Environmental Disasters in the Third World.* Stockholm and Geneva: The Swedish Red Cross, 1984.

Hall, A. D. Forest Fire Prevention in Canada: An Assessment. *Paper and Pulp Magazine of Canada, 70*(21) (1969), 125–127.

Hankins, T. D. Response to Drought in Sukumaland, Tanzania. In G. F. White, ed., *Natural Hazards: Local, National, Global.* New York: Oxford Universtiy Press, 1974, pp. 98–104.

Haque, C. E. and D. Blair. Vulnerability to Tropical Cyclones: Evidence from the April 1991 Cyclone in Coastal Bangladesh. *Disasters, 16* (1992), 217–229.

Hardin, G. The Tragedy of the Commons. *Science, 162* (1968), 1243–1248.

Hardin, G. Nobody Ever Dies of Overpopulation. *Science, 171* (1971), 527.

Harris, R. C., C. Hohenemser, and R. W. Kates. Human and Nonhuman Mortality. In R. W. Kates, C. Hohenemser, and J. X. Kasperson, eds., *Perilous Progress: Managing the Hazards of Technology.* Boulder, CO: Westview Press, 1985, pp. 129–155.

Hattis, D. R., R. Goble, and N. Ashford. Airborne Lead: A Clearcut Case of Differential Protection. *Environment, 24*(1) (1982), 14–20, 33–42.

Heathcote, R. L. Drought in Australia: A Problem of Perception. *Geographical Review, 59* (1969a), 175–194.

Heathcote, R. L. Drought in South Australia. In G. F. White, ed., *Natural Hazards: Local, National, Global.* New York: Oxford University Press, 1974, pp. 128–136.

Heathcote, R. L. The Pastoral Ethic: A Comparative Study of Pastoral Resource Appraisals in Australia and America. In W. G. McGinnies and G. J. Goldman, eds., *Arid Lands in Perspective.* Tucson: University of Arizona Press, and Washington, DC: American Association for the Advancement of Science, 1969b, pp. 312–324.

Heathcote, R. L. Extreme Event Analysis. In R. W. Kates, J. H. Ausubel, and M. Berberian, eds. *Climate Impact Assessment.* ICSU/SCOPE Report 27. Chichester, England: John Wiley, 1985, pp. 369–402

Heberlein, T. A. *Moral Norms, Threatened Sanctions, and Littering Behavior.* Ph.D. dissertation, University of Wisconsin, 1971.

Hecht, S. B., A. B. Anderson, and P. May. The Subsidy from Nature: Shifting Cultivation, Successional Palm Forests, and Rural Development. *Human Organization, 47*(1) (1988), 25–35.

Heijnen, J., and R. W. Kates. Northeast Tanzania: Comparative Observations Along a Moisture Gradient. In G. F. White, ed., *Natural Hazards: Local, National, Global.* New York: Oxford University Press, 1974, pp. 105–114.

Hewapathirane, D. *Flood Hazard in Sri Lanka: Human Adjustments and Alternatives.* Ph.D. dissertation, University of Colorado, Department of Geography, 1977.

Hewitt, K., ed. *Interpretations of Calamity: From the Viewpoint of Ecology.* London: Allen and Unwin, 1983.

Hewitt, K., and I. Burton. *The Hazardousness of a Place: A Regional Ecology of Damaging Events.* Department of Geography Research Paper No. 6, University of Toronto Press, 1971.

Heyman, B. N., C. Davis, and P. F. Krumpe. An Assessment of Worldwide Disaster Vulnerability. *Disaster Management, 4*(1) (1991), 3–14.

Hohenemser, C., R. W. Kates, and P. Slovic. The Nature of Technological Hazard. *Science, 220*(No. 4495) (1983), 378–384.

Hohenemser, C., R. E. Kasperson, and R. W. Kates. Causal Structure. In R. W. Kates, C. Hohenemser, and J. X. Kasperson eds., *Perilous Progress: Managing the Hazards of Technology.* Boulder, CO: Westview Press, 1985, pp. 25–42.

Holling, C. S., D. D. Jones, and W. C. Clark. *Ecological Policy Design: Lessons from a Study of Forest/Pest Management.* Vancouver, BC: University of British Columbia, Institute of Resource Ecology, 1976.

Horlick-Jones, T., and G. Peters. Measuring Disaster Trends. *Disaster Management, 3*(3) (1991), 144–147; *4*(1), 41–44.

Houghton, J. T., G. J. Jenkins, and J. J. Ephramus. *Climate Change: The IPCC Scientific Assessment.* Cambridge, England: Cambridge University Press, 1990.

Housner, G. W. An International Decade for Natural Disaster Reduction, 1990–2000. *Natural Hazards, 2*(1) (1989), 45–75.

Howard, R. A., J. E. Matheson, and D. W. North. The Decision to Seed Hurricanes. *Science, 176* (June 16, 1972), 1191–1202.

Howe, C. W. *Benefit–Cost Analysis for Water System Planning.* Washington, DC: American Geophysical Union, 1971.

Howe, C. W. Savings Recommendations with Regard to Water: System Losses. *Journal of the American Water Works Association, 63* (1971), 284–286.

Howe, C. W., H. C. Cochrane, J. E. Bunin, and R. W. Kling. *Natural Hazard Damage Handbook: A Guide to the Uniform Definition, Identification, and Measurement of Damages from Natural Hazard Events.* Boulder, CO: University of Colorado, Institute of Behavioral Science, Environment & Behavior Program, 1991.

Howe, G. M. The Environment: Its Influences and Hazards to Health. In G. M. Howe, ed., *Environmental Medicine*. London: Heinemann Medical Books, 1980.

Hurst, H. E. *The Nile*. London: Constable, 1952.

Hyman, R., and E. Z. Vogt. Water Witching: Magical Ritual in Contemporary United States. In J. H. Sims and D. D. Baumann, eds., *Human Behavior and the Environment*. Chicago: Maaroufa Press, 1974, pp. 326–344.

Islam, M. A. *Human Adjustment to Cyclone Hazards: A Case Study of Char Jabbar*. Natural Hazards Research Working Paper No. 18. Toronto: University of Toronto, 1971.

Islam, M. A. Tropical Cyclones, Coastal Bangladesh. In G. F. White, ed., *Natural Hazards: Local, National, Global*. New York: Oxford University Press, 1974, pp. 19–24.

Jackson, R. H. Frost Hazard to Tree Crops in the Wasatch Front: Perception and Adjustments. In G. F. White, ed., *Natural Hazards: Local, National, Global*. New York: Oxford University Press, 1974, pp. 146–151.

Jacobson, J. K., and M. F. Price, for the ISSC Standing Committee on the Human Dimensions of Global Change. *A Framework for Research on the Human Dimensions of Global Environmental Change*. Paris: International Social Science Council, 1991.

Johnson, D. L., and H. Gould. The Effects of Climate Fluctuations on Human Populations: A Case Study of Mesopotamian Society. In A. K. Biswas, ed., *Climate and Development*. Dublin: Tycooly International, 1984.

Johnson, K. Publics, Workers, and Parathion: Equity in Hazard Management. *Environment, 24*(9) (1982), 28–38.

Kahneman, D., and A. Tversky. Subjective Probability: A Judgment of Representativeness. *Cognitive Psychology, 3* (1973), 430–454.

Kasperson, R. E. Political Behavior and the Decision Making Process in the Allocation of Water Resources Between Recreational and Municipal Use. *Natural Resources Journal, 8* (1969), 176–211.

Kasperson, R. E., ed. *Equity Issues in Radioactive Waste Management*. Cambridge, MA: Oelgeschlager, Gunn and Hain, 1983.

Kasperson, R. E., P. Derr, and R. W. Kates. Confronting Equity in Radioactive Waste Management: Modest Proposals for a Socially Just and Acceptable Program. In R. E. Kasperson, ed., *Equity Issues in Radioactive Waste Management*. Cambridge, MA: Oelgeschlager, Gunn, and Hain, 1983, pp. 331–368. (Also in M. J. Pasqualetti and K. D. Pijawka, eds., *Nuclear Power: Assessing and Managing Hazardous Technology*. Boulder, CO: Westview Press, 1984, pp. 349–385.)

Kasperson, R. E., R. W. Kates, and C. Hohenemser. Hazard Management. In R. W. Kates, C. Hohenemser, and J. X. Kasperson, eds., *Perilous Progress: Managing the Hazards of Technology*. Boulder, CO: Westview Press, 1985, pp. 43–66.

Kasperson, R. E., and K. D. Pijawka. Societal Response to Hazards and Major Hazard Events: Comparing Natural and Technological Hazards. *Public Administration Review, 45*(1985), 7–18.

Kasperson, R. E., O. Renn, P. Slovic, H. S. Brown, J. Emel, R. Goble, J. X. Kasperson, and S. Ratick. The Social Amplification of Risk: A Conceptual Framework. *Risk Analysis, 8*(2) (1988), 177–185.

Kasperson, R. E. and J. X. Kasperson. Hidden Hazards, In D. G. Mayo and R. D. Hollander, eds., *Acceptable Evidence: Science and Values in Risk Management*.

New York: Oxford University Press, 1991, pp. 9–28.

Kates, R. W. *Hazard and Choice Perception in Flood Plain Management.* Research Paper No. 78, Chicago: University of Chicago, Department of Geography, 1962.

Kates, R. W. *Industrial Flood Losses: Damage Estimation in the Lehigh Valley.* Research Paper No. 98, Chicago: University of Chicago, Department of Geography, 1965.

Kates, R. W. Planning for Hazards in Everyday Landscapes. *Landscape Architecture* (April 1975), *65,* 165–168.

Kates, R. W. Natural Hazard in Human Ecological Perspective: Hypotheses and Models. *Economic Geography, 47* (July 1971), 438–451. Published also as *Natural Hazards Research Working Paper No. 14. Worcester, MA, Clark University, 1970.*

Kates, R. W. *Natural Disasters and Development.* Paper presented at Wingspread Conference, Racine, WI, October 1975.

Kates, R. W. Assessing the Assessors: The Art and Ideology of Risk Assessment. *Ambio 6*(5) (1977a), 247–252.

Kates, R. W., ed. *Managing Technological Hazard: Research Needs and Opportunities.* Environment and Behavior Monograph No. 25. Boulder, CO: University of Colorado, Institute of Behavioral Science, 1977b.

Kates, R. W. *Risk Assessment of Environmental Hazard.* ICSU/SCOPE Report No. 8. Chichester, England: John Wiley & Sons, 1978.

Kates, R. W. Disaster Reduction: Links between Disaster and Development. In L. Berry and R. W. Kates, eds., *Making the Most of the Least: Alternative Ways to Development.* New York: Holmes and Meier, 1980, pp. 135–169.

Kates, R. W. Drought in the Sahel: Competing Views as to What Really Happened in 1910–1914 and 1968–1974. *Mazingira, 5*(2) (1981), 72–83.

Kates, R. W. The Interaction of Climate and Society. In R. W. Kates, J. H. Ausubel, and M. Berberian. ICSU/SCOPE Report 27. Chichester, England: John Wiley, 1985a, pp. 3–36.

Kates, R. W. Success, Strain, and Surprise. *Issues in Science and Technology, 2*(1) (1985b). (Later published as Managing Technological Hazards: Success, Strain, and Surprise. In National Academy of Engineering, Series on Technology and Social Priorities. *Hazards: Technology and Fairness.* Washington, DC: National Academy Press, 1986.)

Kates, R. W. Hazard Assessment and Management. In D. S. McLaren and B. J. Skinner, eds., *Resources and World Development.* Chichester, England: John Wiley, 1987, pp. 741–752.

Kates, R. W. Theories of Nature, Society, and Technology. In Erik Baark and Uno Svedin, eds., *Man, Nature, and Technology: Essays on the Role of Ideological Perceptions.* Houndmills, England: Macmillan Press, 1988, pp. 7–36.

Kates, R. W., J. H. Ausubel, and M. Berberian, eds. *Climate Impact Assessment: Studies of the Interaction of Climate and Society.* ICSU/SCOPE Report 27. Chichester, England: John Wiley & Sons, 1985.

Kates, R. W., W. C. Clark, V. Norberg-Bohm, and B. L. Turner II. *Human Sources of Global Change: A Report on Priority Research Initiatives for 1990–1995.* Occasional Paper No. 3. Providence, RI: Brown University, Institute for International Studies, 1990.

Kates, R. W., W. C. Clark, and B. L. Turner II. The Great Transformation. In B. L. Turner II, W. C. Clark, R. W. Kates, J. F. Richards, J. T. Matthews, and W. B. Meyer, *The Earth as Transformed by Human Action.* Cambridge,

England: Cambridge University Press, 1990, pp. 1–17.

Kates, R. W., and V. Haarmaan. Where the Poor Live: Are the Assumptions Correct. *Environment, 34*(4) (1992), 4–11, 25–28.

Kates, R. W., J. E. Haas. Human Impact of the Managua Earthquake. *Science, 182* (1973), 981–990.

Kates, R. W., C. Hohenemser, and J. X. Kasperson, eds. *Perilous Progress: Managing the Hazards of Technology.* Boulder, CO: Westview Press, 1985.

Kates, R. W., D. L. Johnson, and K. Johnson-Haring. Population, Society, and Desertification. In Secretariat of the United Nations Conference on Desertification, *Desertification: Its Causes and Consequences.* Oxford, England: Pergamon Press, 1977, pp. 261–318.

Kates, R. W., and J. X. Kasperson. Comparative Risk Analysis of Technological Hazards (A Review). *Proceedings of the National Academy of Sciences, U.S.A., 80* (1983), 7027–7038.

Kelly, G. A. *The Psychology of Personal Constructs.* New York: W. W. Norton, 1955.

Kirby, A., ed. *Nothing to Fear: Risks and Hazards in American Society.* Tucson: University of Arizona Press, 1990.

Kirkby, A. V. Individual and Community Response to Rainfall Variability in Oaxaca, Mexico. In G. F. White, ed., *Natural Hazards: Local, National, Global.* New York: Oxford University Press, 1974, pp. 119–128.

Krutilla, J. V., and O. Eckstein. Multiple Purpose River Development. Studies in Applied Economic Analysis. Baltimore: Johns Hopkins University Press, 1958.

Kunreuther, H., Economic Analysis of Natural Hazards: An Ordered Choice Approach. In G. F. White, ed., *Natural Hazards: Local, National, Global.* New York: Oxford University Press, 1974, pp. 206–214.

Kunreuther, H., R. Ginsberg, Miller, P. Sagi, P. Slovic, B. Borkan, and N. Katz, *Disaster Insurance Protection: Public Policy Lessons.* New York: John Wiley & Sons, 1978.

Lawless, E. T. *Technology and Social Shock.* New Brunswick, NJ: Rutgers University Press, 1977.

Lemieux, G. *Study of Volcano Hazard in Costa Rica Area.* Calgary, Alberta: University of Calgary, Department of Geography, 1972. Unpublished.

Leonard, H. J., ed. *Environment and the Poor: Development Strategies for a Common Agenda.* U.S.–Third World Policy Perspectives No. 11. Washington, DC: Overseas Development Council, 1989.

Leopold, L. B., and T. Maddock. *The Flood Control Controversy: Big Dams, Little Dams and Land Management.* New York: Ronald Press, 1954.

Lindblom, C. E., and D. Braybrooke. *A Strategy of Decision.* New York: Free Press, 1963.

Little, P. D., and M. M. Horowitz, eds. *Lands at Risk in the Third World: Local-Level Perspectives.* Boulder, CO: Westview Press, 1987.

Liverman, D. M. Drought Impacts in Mexico: Climate, Agriculture, Technology and Land Tenure in Sonora and Puebla. *Annals of the Association of American Geographers, 80*(1) (1990), 49–72.

Liverman, D. M. The Regional Impacts of Global Warming in Mexico: Uncertainty, Vulnerability and Response. In J. Schmandt, and J. Clarkson, ed., *The Regions and Global Warming.* Cambridge, England: Cambridge University Press, 1992.

Liverman, D. M., W. H. Terjung, J. T. Hayes, and L. O. Mearns. Climatic

Change and Grain Corn Yields in the North American Great Plains. *Climatic Change, 9*(1986), 327–347.

Lowrence, W. W. *Of Acceptable Risk: Science and the Determination of Safety.* Los Altos, CA: William Kaufmann, 1976.

Lutz, H. J., Aboriginal Man and White Man as Historical Causes of Fires in the Boreal Forest, with Particular Reference to Alaska. *Yale University School of Forestry Bulletin, 65*(1959), pp. 1–49.

Malone, T. Mission to Planet Earth: Integrating Studies of Global Change, *Environment, 28*(1986), 8–15, 30–31.

Mbithi, P. M., and B. Wisner. Drought in Eastern Kenya: Nutritional Status and Farmer Activity. In G. F. White, ed., *Natural Hazards: Local, National, Global.* New York: Oxford University Press, 1974, pp. 87–97.

McKean, R. N. *Efficiency in Government Through Systems Analysis with Emphasis on Water Resources Development.* New York: Wiley, 1958.

McKell, C. Review of the 1977–1987 Decade of Action to Combat Desertification. *Population and Environment, 11*(1) (1989), 25–30.

Melville, M. *The Temporary Worker in the Nuclear Power Industry: An Equity Analysis.* Monograph No. 1. Worcester, MA: Clark University Center for Technology, Environment, and Development, 1981.

Metropolitan Toronto and Region Conservation Authority. *Plan for Flood Control and Water Conservation.* Woodbridge, Ontario: The Authority, 1959.

Mileti, D. S. Natural Hazard Warning Systems in the United States: A Research Assessment. Monograph No. NSF-RA-E-75-013, Boulder: University of Colorado, Institute of Behavioral Science, Program on Technology, Environment and Man, 1975.

Mileti, D. S., B. C. Farhar and C. Fitzpatrick. *How to Issue and Manage Public Earthquake Risk Information: Lessons from the Parkfield Earthquake Prediction Experiment.* Fort Collins, CO: Colorado State University, Hazards Assessment Laboratory, 1990.

Mileti, D. S., C. Fitzpatrick and B. C. Farhar. *Risk Communication and Public Response to the Parkfield Earthquake Prediction Experiment.* Fort Collins, CO: Colorado State University, Hazards Assessment Laboratory and Department of Sociology, 1990, p. 200.

Millman, S., and R. W. Kates. Toward Understanding Hunger. In L. F. Newman, W. Cosgrove, R. W. Kates, R. Mathews, and S. Millman, eds., *Hunger in History: Food Shortage, Poverty, and Deprivation.* Oxford: Basil Blackwell, 1990, 3–24.

Mitchell, J. K. *Community Response to Coastal Erosion.* Research Paper No. 156. Chicago: University of Chicago, Department of Geography, 1974.

Mitchell, J. K. Adjustment to New Physical Environments Beyond the Metropolitan Fringe. *Geographical Review, 66* (January 1976), 18–31.

Mitchell, J. K. Confronting Natural Disasters: An International Decade for Natural Hazard Reduction. *Environment, 30*(2) (1988), 25–29.

Mitchell, J. K. Hazards Research. In G. L. Gaile and C. J. Willmott, eds., *Geography in America.* Columbus, OH: Merrill, 1989, pp. 410–424.

Mitchell, J. K. *Risk Assessment of Global Environmental Change.* Working Paper No. 13. Honolulu: East-West Center, 1989.

Mittler, E. *Building Code Enforcement Following Hurricane Hugo in South Carolina.* Quick Response Report No. 44. Boulder, CO: University of Colorado, Natural Hazards Research and Applications Information Center, 1991.

Moline, N. T. Perception Research and Local Planning: Floods on the Rock River, Illinois. In G. F. White, ed., *Natural Hazards: Local, National, Global*. New York: Oxford University Press, 1974, pp. 52–59.

Moon, K. D. *The Perception of the Hazardousness of a Place: A Comparative Study of Five Natural Hazards in London, Ontario*. M.A. research paper, University of Toronto, 1971.

More, R. J. M. *A Geographical Analysis of Irrigation Use in South East England, with Particular Reference to the Great Ouse Valley*. Ph.D. dissertation, University of Liverpool, 1964.

Munroe. T. and K. P. Ballard. Modeling the Economic Disruption of a Major Earthquake in the San Francisco Bay Area: Impact on California. *Annals of Regional Science, 17*(3) (1983): 23–40.

Murton, B. J., and S. Shimabukuro. Human Adjustment to Volcanic Hazard in Puna District, Hawaii. In G. F. White, ed., *Natural Hazards: Local, National, Global*. New York: Oxford University Press, 1974, pp. 151–159.

Mutch, R. W. Wildland Fires and Ecosystems: A Hypothesis. *Ecology, 51*(6) (1970), 1046–1051.

Myrdal, J., ed. *Report from a Chinese Village*. London: William Heinemann, 1963.

Nakano, T., et al. Natural Hazards: Report from Japan. In G. F. White, ed., *Natural Hazards: Local, National, Global*. New York: Oxford University Press, 1974, pp. 231–242.

National Academy of Sciences, International Environmental Programs Committee. *Early Action on the Global Environmental Monitoring System*. Washington, DC: National Academy of Sciences, 1976.

National Research Council. *Confronting Natural Disasters: An International Decade for Natural Hazard Reduction*. Washington, DC: National Academy Press, 1987.

National Research Council. *Managing Coastal Erosion*. Washington, DC: National Academy Press, 1990.

Natural Hazards Research and Applications Information Center. *Report of the Colorado Workshop on Hazard Mitigation in the 1990s*. Boulder, CO: University of Colorado, Institute of Behavioral Science, 1989.

Ngugi, J. *Secret Lives and Other Stories*. New York: Lawrence Hill and Co., 1975.

Nordhaus, W. To Slow or not to Slow: The economics of the Greenhouse Effect. *Economic Journal, 101*(407) (1991), 920–937.

O'Keefe, P., B. Wisner, and A. Baird. Kenyan Underdevelopment: A Case Study of Proletarianisation. In P. O'Keefe and B. Wisner, eds., *Landuse and Development*. African Environment Special Report No. 5. London: International African Institute, 1977, pp. 216–228.

O'Riordan, T. *The New Zealand Earthquake and War Damage Commission — A Study of a National Natural Hazard Insurance Scheme*. National Hazards Research Working Paper No. 20. Toronto: University of Toronto, 1971a.

O'Riordan, T. Public Opinion and Environmental Quality: A Reappraisal. *Environment and Behavior, 3* (1971b), 191–214.

O'Riordan, T. The New Zealand Natural Hazard Insurance Scheme: Application to North America. In G. F. White, ed., *Natural Hazards: Local, National, Global*. New York: Oxford University Press, 1974, pp. 217–219.

O'Riordan, T. Coping with Environmental Hazards. In R. W. Kates and I. Burton. eds., *Geography, Resources, and Environment*, Vol. 2, 1986, pp. 272–309. Chicago: University of Chicago.

Page, R. A., and W. B. Joyner. *Probable Shaking of San Francisco Bay Area in the 1906*

Type Earthquake. Paper presented at 14th meeting of the American Association for the Advancement of Science, February 27, 1974.

Palm, R. I. *Natural Hazards: An Integrative Framework for Research and Planning*. Baltimore: Johns Hopkins University Press, 1990.

Parra, C. G. *Drought Perception and Adjustments in Yucatan, Mexico*. M.A. thesis, University of Colorado, 1971.

Parry, M. L. The Impact of Climatic Variations on Agricultural Margins, In R. W. Kates, J. H. Ausubel, and M. Berberian, eds., *Climate Impact Assessment*. ICSU/SCOPE Report 27. Chichester, England: John Wiley & Sons, 1985, pp. 351–368.

Parry, M. L. *Climate Change and World Agriculture*. London: Earthscan Publications, 1990.

Parry, M. L., T. R. Carter, and N. T. Konjin, eds. *The Impacts of Climatic Variations on Agriculture*, Vol. 1; *Assessments in Cool and Temperate Regions*; Vol. 2: *Assessments in Semi-Arid Regions*. Hingham, MA: Kluwer, 1988.

Penaherra de Aguila, C. *Study of Avalanche Hazard in Yungay, Peru Area*. Lima, Peru: University of San Marcos and National Planning Institute, 1970. Unpublished.

Pijawka, K. D., B. A. Cuthbertson, and R. S. Olson. Coping with Extreme Hazard Events. *Omega, 18*(4) (1988), 281–297.

Pinchot, G. *Breaking New Ground*. New York: Harcourt, Brace, 1947.

Porter, P. The Concept of Environmental Potential as Exemplified by Tropical African Research. In W. Zelinsky, L. A. Kosinski, and M. Prothero, ed., *Geography and a Crowding World*, New York: Oxford University Press, 1970, pp. 187–217.

Pushchak, R. and I. Burton. Risk and Prior Compensation in Siting Low-level Waste Facilities: Dealing with the NIMBY Syndrome. *Plan Canada, 23(3)* (1983), 68–79.

Rahman, A. Disaster and Development: A Study in Institution Building in Bangladesh. Paper presented at the UCLA International Conference on the Impact of Natural Disasters, Los Angeles, CA; July 10–12, 1991.

Renshaw, E. F. *Toward Responsible Government: An Economic Appraisal of Federal Investment in Water Resource Programs*. Chicago: Indyia Press, 1957.

Richards, P. *Coping with Hunger: Hazard and Experiment in an African Rice-Farming System*. London: Allen and Unwin, 1986.

Riebsame, W. E., H. F. Diaz, T. Moses, and M. Price. The Social Burden of Weather and Climate Hazards. *Bulletin American Meteorological Society, 67*(11) (1986), 1378–1388.

Riebsame, W. E., S. A. Changnon, Jr., and T. R. Karl. *Drought and Natural Resources Management in the United States: Impacts and Implications of the 1987–1989 Drought*. Boulder, CO: Westview Press, 1991.

Rivers, J. P. W. Women and Children Last: An Essay on Sex Discrimination in Disasters. *Disasters, 6*(4) (1982), 256–267.

Rooney, J. The Economic and Social Implications of Snow and Ice. In R. J. Chorley, ed., *Water, Earth and Man*, London: Methuen, 1969, pp. 389–401.

Rosenberg, N. J., and P. R. Crosson. *Processes for Identifying Regional Influences and Responses to Increasing Atmospheric CO_2 and Climate Change — The MINK Project: An Overview*. Office of Energy Research Report TR052A. Washington, DC: U.S. Department of Energy, 1991.

Rowntree, R. A. Coastal Erosion: The Meaning of a Natural Hazard in the Cultural and Ecological Context. In G. F. White, ed., *Natural Hazards: Local, National, Global*. New York: Oxford University Press, 1974, pp. 70–79.

Rubin, C. B. and R. Popkin. *Disaster Recovery After Hurricane Hugo in South Carolina*. Working Paper No. 69. Boulder, CO: University of Colorado, Natural Hazards Research and Applications Information Center, 1991.

Russell, C. S., D. G. Arey, and R. W. Kates. *Drought and Water Supply. Implications of the Massachusetts Experiences for Municipal Planning*. Baltimore: Johns Hopkins University Press, 1970.

Saarinen, T. F. *Perception of Drought Hazard on the Great Plains*. Research Paper No. 106. Chicago: University of Chicago, Department of Geography, 1966.

Sandberg, A. *Ujamaa and Control of Environment*. Paper presented at the Annual Social Science Conference of East African Universities, December 1973.

Sandberg, A. Socio-Economic Survey of Lower Rufiji Flood Plain. Research Paper No. 34. Dar es Salaam, Tanzania: University of Dar es Salaam, Bureau of Land Planning, October 1974.

Sauer, C. O. A Geographic Sketch of Early Man in America. *Geographical Review, 34* (1944): 529–573.

Sauer, C. O. Grassland Climax, Fire, and Man. *Journal of Range Management, 3* (1950), 16–21.

Scientific Committee on Problems of the Environment. *Comparative Risk Assessment*. Report of Workshop on Comparative Risk Assessment of Environmental Hazards in an International Context. Mid-Term Project 7. Woods Hole, MA: SCOPE Miscellaneous Publications, April 1975.

Scott, E., ed. *Life before the Drought*. Boston: Allen and Unwin, 1984.

Scudder, T. Social Anthropology, Man-Made Lakes and Population Relocation in Africa. *Anthropological Quarterly, 41* (1968), 168–176.

Selye, H. *The Stress of Life*. New York: McGraw Hill, 1956.

Selye, H. *Stress without Distress*. Philadelphia, PA: Lippincott, 1974.

Sen, A. K. *Poverty and Famine: An Essay on Entitlement and Deprivation*. Oxford: Clarendon Press, 1982.

Sewell, W. R. D. *Water Management and Floods in the Fraser River Basin*. Research Paper No. 100. Chicago: University of Chicago, Department of Geography, 1965.

Shaeffer, J. R., ed. *Introduction to Flood Proofing*. Chicago: University of Chicago Press, 1967.

Sheehan, L., and K. Hewitt. *A Pilot Survey of Global Natural Disasters of the Past Twenty Years*. Natural Hazards Research Working Paper No. 11. Toronto: University of Toronto, 1969.

Showalter, P. S. *Field Observations in Memphis during the New Madrid Earthquake "Projection" of 1990: How Pseudoscience Affected a Region*. Natural Hazards Research Working Paper No. 71. Boulder, CO: University of Colorado, Institute of Behavioral Science. 1991.

Showalter, P. S., W. E. Riebsame, and M. F. Myers, *Natural Hazard Trends in the United States: A Preliminary Review for the 1990s*. Natural Hazards Research Working Paper No. 82. Boulder, CO: University of Colorado, Institute of Behavioral Science, 1993.

Simkowski, N. A. *The Structure of Influence in Adoption of Flood Plain Regulations*. M.A. thesis, University of Colorado, 1973.

Simon, H. A. Rational Choice and the Structure of the Environment. *Psychological Review, 63* (1956), 129–138.

Slovic, P., B. Fischhoff, and S. Lichtenstein. Cognitive Processes and Societal Risk Taking. In M. Carroll and J. Payne, eds., *Cognition and Social Behavior*. Potomac, MD: Lawrence Erlbaum Associates, 1976.

Slovic, P., B. Fischhoff, and S. Lichtenstein. Facts versus Fears: Understanding Perceived Risk. In D. Kahneman, P. Slovic, and A. Tversky, eds., *Judgement Under Uncertainty: Heuristics and Biases*. Cambridge, England: Cambridge University Press, 1982, pp. 463–489.

Slovic, P., B. Fischhoff, and S. Lichtenstein. Characterizing Perceived Risk. In R. W. Kates, C. Hohenemser, and J. X. Kasperson, eds., *Perilous Progress*. Boulder, CO: Westview Press, 1985, pp. 91–125.

Slovic, P., H. Kunreuther, and G. F. White. Decision Processes, Rationality, and Adjustment to Natural Hazards. In G. F. White, ed., *Natural Hazards: Local, National, Global*. New York: Oxford University Press, 1974, pp. 187–205.

Smith, K. *Environmental Hazards: Assessing Risk and Reducing Disaster*. London: Routledge, 1992.

Starr, C. Benefit–Cost Studies in Sociotechnical Systems. In *Perspectives on Benefit–Risk Decision Making*. Washington, DC: National Academy of Engineering, 1972, pp. 17–42.

Stern, P. C., O. R. Young, and D. Druckman, eds. *Global Environmental Change: Understanding the Human Dimensions*. Washington, DC: National Academy Press, 1992.

Stewart, O. S. Fire as the First Great Force Employed by Man. In W. L. Thomas, Jr., ed., *Man's Role in Changing the Face of the Earth*. Chicago: University of Chicago Press, 1956, pp. 115–133.

Stewart, O. C. *Barriers to Understanding the Influence of Use of Fire by Aborigines on Vegetation*. Report from 2nd Annual Proceedings, Tall Timbers Fire Ecology Conference, Tallahassee, FL, 1963, pp. 117–126.

Sugg, A. E. Economic Aspects of Hurricanes. *Monthly Weather Review, 95* (March 1967), 143–146.

Susman, P., P. O'Keefe, and B. Wisner. Global Disasters, a Radical Interpretation. In K. Hewitt, ed., *Interpretation of Calamity*, 1983, London: Allen and Unwin, pp. 263–283.

Tait, J., ed. *Perception and Management of Pests and Pesticides: Guidelines for Research*. Working Paper EPR-8. Toronto: University of Toronto, Institute for Environmental Studies, 1981.

Tait, J., and B. Napompeth, eds. *Management of Pests and Pesticides: Farmer's Perceptions and Practices*. Boulder, CO: Westview Press, 1987.

Thompson, S. A. *Trends and Developments in Global Natural Disasters, 1947 to 1981*. Natural Hazards Research and Applications Information Center Working Paper No. 41. Boulder, CO: University of Colorado, Institute of Behavioral Science, 1982.

Tokyo Earthquake Casualties. *Tokyo Municipal News*, December 1973.

Torry, W. I. Hazards, Hazes, and Holes: A Critique of the Environment as Hazard and General Reflections on Disaster Research. *Canadian Geographer, 23* (1979), 368–383.

Torry, W. I. Economic Development, Drought, and Famines: Some Limitations of Dependence Explanations. *GeoJournal, 12*(1) (1986), 5–14.

Townley, R., The United Nations: A View from Within. New York: Scribner, 1968.

Turner, B. L. II, W. C. Clark, R. W. Kates, J. F. Richards, J. T. Mathews, and W. B. Meyer, eds. *The Earth as Transformed by Human Action: Global and Regional Changes in the Biosphere over the Past 300 Years.* Cambridge, England: Cambridge University Press, 1990.

Tversky, A., and D. Kahneman. Belief in the Law of Small Numbers. *Psychological Bulletin, 76* (1971), 105–110.

United Kingdom Climate Change Impacts Review Group. *The Potential Effects of Climate Change in the United Kingdom.* London: Her Majesty's Stationery Office, 1991.

UNESCO. *Expert Panel on Project 13: Perception of Environmental Quality, Final Report.* Paris: Programme on Man and the Biosphere (MAB). Report No. 9, March 26–29, 1973.

United Nations Department of Economics and Social Affairs. *The Role of Science and Technology in Reducing the Impacts of Natural Disasters on Mankind.* Report of Advisory Committee on the Applications of Science and Technology to Development, New York, 1972.

United Nations General Assembly. *Report of the United Nations Conference on the Human Environment.* No. A/Conf. 48/14. Stockholm, June 5–6, 1972.

United Nations General Assembly. Resolution 42/169 (International Decade for Natural Disaster Reduction) adopted December 11, 1987. U.N. General Assembly 1987a. Reprinted in *UNDRO News,* 2, (January/February 1988), 24–25.

United Nations Disaster Relief Office. Tragic Cyclone in Bangladesh. *UNDRO News, 5,* (March–April 1991), 5–7.

United Nations General Assembly, Intergovernmental Negotiating Committee for a Framework Convention on Climate Change. *Report of the Intergovernmental Negotiatig Committee for a Framework Convention on Climate Change on the Work of the Second Part of its fifth session, held at New York frm 30 April to 9 May 1992.* New York: United Nations, 1992.

United States, Congress. *A Unified National Program for Managing Flood Losses.* 87th Congress, 2nd session, 1966, House Document 465. Washington, DC: Government Printing Office, 1966.

United States Department of Commerce. *Hurricane Hugo, September 10–22, 1989.* Silver Spring, MD: Department of Commerce, National Oceanic and Atmospheric Administration, National Weather Service, Natural Disaster Survey Report, 1990.

United States Department of Defense. *Hurricane Hugo: After-Action Report.* Charleston, SC: United States Department of Defense, Army Corps of Engineers, Charleston District, 1990.

United States Federal Emergency Management Agency. *Hurricane Hugo: Interagency Hazard Mitigation Team Report.* Report No. FEMA-843-DR-SC. Atlanta, Georgia: U.S. Federal Emergency Management Agency, 1989.

United States Federal Interagency Floodplain Management Task Force. *Floodplain Management in the United States: An Assessment Report*—Volume 1: *Summary.* FIA-17. Washington, DC: U.S. Federal Emergency Management Agency, 1992a.

United States Federal Interagency Floodplain Management Task Force. *Floodplain*

Management in the United States: An Assessment Report—Volume 2: *Full Report.* FIA-18, Washington, DC: U.S. Federal Emergency Management Agency, 1992b.

United States General Accounting Office. *Disaster Assistance: Federal, State, and Local Responses to Natural Disasters Needs Improvement.* Report No. GAO/RCED-91-43. Washington, DC: U.S. General Accounting Office, 1991.

United States Office of Emergency Preparedness. *Disaster Preparedness: Report to the Congress.* Washington, DC: Government Printing Office, 1972.

United States Office of Foreign Assistance. *Disaster History: Significant Data on Major Disasters Worldwide, 1900–Present.* Washington, DC: Office of U.S. Foreign Disaster Assistance, Agency for International Development, 1991.

United States Weather Service. United States Department of Commerce, National Weather Service, Miami National Hurricane Center Technical Memorandum No. 31, 1990.

Velimerovic, H. *An Anthropological View of Risk Phenomena.* Research Memorandum. Schloss Laxenburg, Austria: International Institute for Applied Systems Analysis, November 1975.

Visvader, H., and I. Burton. Natural Hazards and Hazard Policy in Canada and the United States. In G. F. White, ed., *Natural Hazards: Local, National, Global.* New York: Oxford University Press, 1974, pp. 219–230.

Waddell, E. The Hazards of Scientism. *Human Ecology,* 5 (1977), 69–76.

Walker, R. Review of I. Burton, et al., *Environment as Hazard. Geographical Review.* 69 (1979), 113–114.

Ward, R. M. Decisions by Florida Citrus Growers and Adjustments to Freeze Hazards. In G. F. White, ed., *Natural Hazards: Local, National, Global.* New York: Oxford University Press, 1974, pp. 137–145.

Warrick, R. A. Drought on the Great Plains: A Case Study of Research on Climate and Society in the U.S.A. In J. Ausubel, and A. Biswas, eds., *Climatic Constraints and Human Activities.* Oxford: Pergamon Press, 1980, pp. 93–123.

Warrick, R. A. The Possible Impacts on Wheat Production of a Recurrence of the 1930s Drought in the U.S. Great Plains. *Climatic Change, 6* (1984), 5–26.

Warrick, R. A., E. M. Barrow, and T. M. L. Wigley, eds. *Climate and Sea Level Change: Observations, Projections, and Implications.* Cambridge, England: Cambridge University Press, 1991.

Warrick, R. A. and M. Bowden. The Changing Impacts of Drought on the Great Plains. In M. P. Lawson, and M. E. Baker, eds. *The Great Plains: Perspectives and Prospects.* Lincoln, NE: Center for Great Plains Studies, 1981, pp. 111–137.

Warrick, R. A., and G. Farmer. The Greenhouse Effect, Climatic Change, and Sea Level: Implications for Development. *Transactions, Institute of British Geographers,* N.S. *15*(1) (1990), 5–20.

Warrick, R. A., and R. M. Gifford, (with M. L. Parry). CO_2, Climatic Change and Agriculture: Assessing the Response of Food Crops to the Direct Effects of Increased CO_2 and Climatic Change. In B. Bolin, B. R. Döös, R. A. Warrick, and J. Jäger, eds., *The Greenhouse Effect, Climatic Change and Ecosystems.* SCOPE No. 29. Chichester, England: John Wiley & Sons, 1986, pp. 393–473.

Watts, M. On the Poverty of Theory: Natural Hazards Research in Context. In Kenneth Hewitt, ed., *Interpretations of Calamity from the Viewpoint of Human Ecology,* 1983a, 231–262.

Watts, M. *Silent Violence: Food, Famine, and Peasantry in Northern Nigeria*. Berkeley: University of California Press, 1983b.

White, G. F. Human Adjustment to Floods: A Geographical Approach to the Flood Problem in the United States. Research Paper No. 29. Chicago: University of Chicago, Department of Geography, 1945.

White, G. F., ed. Papers on Flood Problems. Research Paper No. 70, Chicago: University of Chicago, Department of Geography, 1961.

White, G. F., ed. Natural Hazards: Local, National, Global. New York: Oxford University Press, 1974.

White, G. F. Paths to Risk Analysis. *Risk Analysis, 8*(2) (1988), 171–175.

White, G. F. When May a Post-audit Teach Lessons? In H. Rosen and M. Reuss eds., *The Flood Control Challenge: Past, Present, and Future*. Chicago: Public Works Historical Society, 1988.

White, G. F. Tasks for the Science Community. In Norwegian Research Council for Science and the Humanities. *Sustainable Development, Science, and Policy*. Oslo: Norwegian Research Council for Science and the Humanities, 1990, pp. 25–30.

White, G. F. Greenhouse Gases, Nile Snails, and Human Choice. In R. Jessor, ed., *Perspectives on Behavioral Science: The Colorado Lectures*. Boulder, CO: Westview Press, 1991, pp. 276–305.

White, G. F., and J. E. Haas. *Assessment of Research on Natural Hazards*. Cambridge, MA: M.I.T. Press, 1975.

White, G. F., D. J. Bradley, and A. U. White. *Drawers of Water. Domestic Water Use in East Africa*. Chicago: University of Chicago Press, 1972.

White, G. F., W. C. Calef, J. W. Hudson, H. M. Mayer, J. R. Sheaffer, and D. J. Volk. *Changes in Urban Occupance of Flood Plains in the United States*. Working Paper 57. Chicago: University of Chicago, Department of Geography, 1958.

White, G. F., W. A. R. Brinkmann, H. C. Cochrane, and N. J. Ericksen. *Flood Hazard in the United States: A Research Assessment*. Monograph No. NSF-RA-E-75-006. Boulder: University of Colorado, Institute of Behavioral Science, Program on Technology, Environment, and Man, 1975.

Whitmore, T. M., B. L. Turner II, D. L. Johnson, R. W. Kates, and T. R. Gottschang. Long-term Population Change. In B. L. Turner, II, W. C. Clark, R. W. Kates, J. F. Richards, J. T. Mathews, and W. B. Meyer, eds., *The Earth Transformed by Human Action*. Cambridge, England: Cambridge University Press, 1990, pp. 25–39.

Whyte, A. V. T. The Role of Information Flow in Controlling Industrial Lead Emissions: The Case of the Avonmouth Smelter. In *Proceedings of the International Conference on Heavy Metals*. Toronto: Institute for Environmental Studies, 1977.

Whyte, A. V. T., and I. Burton, eds. *Environmental Risk Assessment*. SCOPE Report No. 15. Chichester, England: John Wiley and Sons, 1980.

Whyte, A. V. T. From Hazard Perception to Human Ecology. In R. W. Kates and I. Burton, eds., *Geography, Resources, and Environment*, Vol. 2. Chicago, University of Chicago Press, 1986, pp. 240–271.

Wigley, T. M. L. and S. C. B. Raper. Implications for Climate and Sea Level of Revised IPCC Emissions Scenarios, *Nature, 387* (1992), 293–300.

Wijkman, A., and L. Timberlake. *Natural Disasters: Acts of God or Acts of Man*. Washington, DC: Earthscan, 1986.

Wisner, B. G., Jr. *The Human Ecology of Drought in Eastern Kenya*. Ph.D. dissertation, Clark University, Worcester, MA, 1977.

Wisner, B. *Power and Need in Africa*. Trenton, NJ: Africa World Press, 1989.

Wittfogel, K. A. *Oriental Despotism: A Comparative Study of Total Power*. New Haven, CT: Yale University Press, 1957.

Wollman, N., and G. W. Bonem. The Outlook for Water: Quality, Quantity, and National Growth. Baltimore: Johns Hopkins University Press, 1971.

Wong, S. T. *Perception of Choice and Factors Affecting Industrial Water Supply Decisions in Northeastern Illinois*. Chicago: University of Chicago, Department of Geography, Working Paper No. 117, 1968.

World Commission on Environment and Development. *Our Common Future*. New York: Oxford University Press, 1987.

Index